THIS BOOK IS DUE ON THE DATE INDICATED BELOW AND IS SUB-JECT TO AN OVERDUE FINE AS POSTED AT THE CIRCULATION DESK.

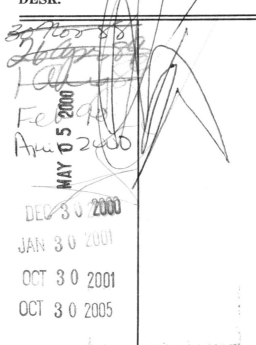

100M/7-87—871203

INTRODUCTION TO STOCHASTIC DIFFERENTIAL EQUATIONS

MONOGRAPHS AND TEXTBOOKS IN
PURE AND APPLIED MATHEMATICS

Other Volumes in Preparation

INTRODUCTION TO STOCHASTIC DIFFERENTIAL EQUATIONS

THOMAS C. GARD

The University of Georgia
Athens, Georgia

MARCEL DEKKER, INC.　　　　　New York and Basel

Library of Congress Cataloging-in-Publication Data

Gard, T. C.
  Introduction to stochastic differential equations / Thomas C.
Gard.
    p. cm. -- (Monographs and textbooks in pure and applied
mathematics ; 114)
  Bibliography: p.
  Includes index.
  ISBN 0-8247-7776-X
  1. Stochastic differential equations. I. Title. II. Series:
Monographs and textbooks in pure and applied mathematics ; v. 114.
QA274.23.G37 1988   87-21278
519.2--dc19   CIP

MARCEL DEKKER, INC.
270 Madison Avenue, New York, New York 10016

Current printing (last digit):
10 9 8 7 6 5 4 3 2 1

PRINTED IN THE UNITED STATES OF AMERICA

To my mother, Ann

and

In memory of my father, John

Preface

During the past forty years, interest in the study of stochastic phenomena has increased dramatically. Intensified research activity in this area has been stimulated by the need to take into account random effects in complicated physical systems. For such systems, stochastic processes, Markov processes in particular, have provided a natural replacement for deterministic functions as mathematical descriptions of state. Consequently, a number of methods for constructing these processes have been introduced. One such method which is well known and extensively applied obtains the distribution of the process from Kolmogorov's equations; however, these partial differential equations for the transition density of the process can be difficult to solve. Another technique of particular relevance when the random effects can be considered as external to some reasonably well-understood deterministic system replaces input parameters in a deterministic model by random processes; the resulting random model characterizes the sample path structure of the solution process. An example which has had wide application in engineering consists of adding a noise term to the right side of a deterministic differential equation. The first such model was formulated by Langevin in 1908 to describe the velocity of a particle moving in a random force field. Random equations of this type can be interpreted as stochastic differential equations, following Ito's basic work in the early 1940s. Solutions of such equations represent Markov diffusion processes, the prototype of which is the Brownian motion (or Wiener) process. The theory of stochastic differential equations set down by Ito, and independently

established in the Soviet Union by Gihman, together with the previous
mathematical work of Wiener and Levy on Brownian motion has provided
the basic tools making this more ambitious approach of constructing sample
paths feasible.

In the past decade applications of stochastic differential equations
have extended beyond engineering and the physical sciences to other areas
including biology and the social sciences. It seems reasonable therefore to
present a basic account of the subject which is accessible to individuals who
do not have extensive graduate training in mathematics, and yet which is as
mathematically rigorous as possible. Recently a number of books treating
stochastic differential equations have appeared. Most of these are either
technically complicated as a result of setting up the stochastic calculus for
the most general class of processes possible or are oriented toward applications in engineering and the physical sciences.

This book, on the other hand, presents the fundamental theory of
stochastic differential equations in the simplest setting so that individuals
having had only introductory courses in real analysis with some ordinary
differential equations theory and probability, including basics of continuous
time stochastic processes, can follow. A concise summary of the probability theory required as background opens the discussion in Chapter 1. Then
Chapters 2 and 3 establish the basic machinery of the Ito stochastic calculus, as the Ito integral is defined and the fundamental existence, uniqueness, and dependence on parameters results for stochastic differential equations are given. Methods for obtaining explicit expressions for solutions of
stochastic differential equations are exhibited in Chapter 4. When explicit
solutions cannot be obtained, one resorts to qualitative and/or quantitative
techniques for obtaining information about solutions. These methods are
discussed in Chapters 5 and 7 with the focus on stability and local error,
respectively. A treatment of applications to population biology is given in
Chapter 6.

The emphasis in this book is on the most important tool of stochastic
analysis—namely, Ito's formula, the stochastic chain rule. It is the key to
understanding the mysteries of stochastic calculus. The last three chapters
demonstrate the central role of Ito's formula in the analytic theory and
applications of stochastic differential equations. It is natural to expect that
the rich theory for ordinary differential equations would provide a source
for an analogous stochastic theory if the stochastic differential equation
terminology is at all meaningful. And, indeed, this is the case. For example,
the important differential inequality-comparison principle technique has a
stochastic analog which is implemented by Ito's formula.

I am indebted to Professor Lou Gross of the University of Tennessee
for his critical review of Chapters 2 through 5, and to Mrs. Cindy Williams

and Mrs. Dianne Byrd for their careful typing of the manuscript and patience in making repeated changes. I also thank Professor Gian-Carlo Rota of the Massachusetts Institute of Technology for his initial solicitation of this monograph. Finally, this work would not have been completed without the constant encouragement of my wife, Carolyn.

Thomas C. Gard

Contents

INTRODUCTION TO
STOCHASTIC
DIFFERENTIAL
EQUATIONS

1

Probability and Stochastic Processes

1.0 INTRODUCTION

The purpose of this chapter is to review the basic facts concerning probability and stochastic processes that are important in understanding stochastic differential equations and their applications. It is expected that the reader is familiar with most of this material, and so the object is to present these topics as concisely as possible while establishing notation. This is done at some expense of completeness—proofs of theorems are omitted, for example. The several good texts currently available (see references) on this subject together with the expediency of getting to the heart of the main topic of discussion seem to warrant this approach.

A solution of an ordinary differential equation is a function defined on some interval whose rate of change at any point in that interval is specified by the differential equation. In a specific application the differential equation constitutes a mathematical model of how a particular physical system evolves during a time interval; the solution gives the state of the system at each point in the interval. Uncertainty in the factors influencing the system's evolution, due to a random environment for example, may be modeled by requiring the differential equation to be stochastic. The solution which describes the corresponding uncertain state of the system is a random or stochastic process. This means that rather than a single function, the solution consists of a family of functions which describe all possible dynamic behaviors of the system. A measure of the relative likelihood of these possibilities is indicated by a probability associated with the family. Mathematically, a probability space has been introduced. At a specific instant in the interval of definition, the solution corresponds to a random variable. Its distribution specifies the level of certainty that one associates with any possible set of values (event) for the solution at that instant. So the discussion opens with a brief treatment of probability spaces, random variables, and distributions, the basis of probability theory.

In the past 50 years, probability has become a legitimate object of study for serious mathematicians. The work of Kolmogorov in the 1930s gave a rigorous mathematical foundation for probability theory by establishing a system of axioms for the subject. This work was, in a way, the natural culmination of over a century of effort by the "Russian school" of probabilists that included such names as Chebyshev, Lyapunov, and Markov and whose principal contributions included providing the first rigorous proofs for the fundamental limit theorems of probability theory—the general central limit theorem and laws of large numbers. Generally, probability theory is considered to have been born with the work on games of chance by Pascal and Fermat at the beginning of the seventeenth century. The enthusiasm resulting from the success of the early work in probability unfortunately led to attempts to "apply" the subject to social, moral, philosophical and political problems. The proliferation of these unjustified "applications" may have acted as a deterrent to serious work by mathematicians, which could explain the approximately 150-year gap in the development of the subject. In any case, now probability occupies a respected place in modern mathematics and is rich in theory and applications.

1.1 PROBABILITY SPACES, RANDOM VARIABLES, AND DISTRIBUTIONS

The principal concept of probability theory is the *probability space*, which can be represented mathematically as an ordered triple $(\Omega, \mathcal{A}, P)$, consisting of the *sample space* Ω, a σ-algebra $\mathcal{A}$ of subsets of Ω called *events*, and a real-valued set function P defined on $\mathcal{A}$ called a *probability*. Ω can be an arbitrary set, whereas $\mathcal{A}$ and P must satisfy the following properties.

The sample space $\Omega \in \mathcal{A}$.   (1.1)

If $A \in \mathcal{A}$, then the complement of A,

$$A^c = \{w \in \Omega : w \notin A\} \in \mathcal{A}. \tag{1.2}$$

For any sequence $\{A_n\} \subseteq \mathcal{A}$, the union $\cup A_n \in \mathcal{A}$.   (1.3)

If $A \in \mathcal{A}, \quad P(A) \geq 0.$ (1.4)

If $\{A_n\} \subseteq \mathcal{A}$, then $P(\cup A_n) = \sum P(A_n)$ whenever

$$A_i \cap A_j = \emptyset \text{ if } i \neq j. \tag{1.5}$$

$$P(\Omega) = 1. \tag{1.6}$$

Properties (1.1) to (1.3) define $\mathcal{A}$ as a *σ-algebra* of subsets of Ω, while properties (1.4) and (1.5) determine that P is a measure, and (1.6) defines P as a *probability (measure)*.

If $A \in \mathcal{A}$, $P(A)$, the probability of A, can be interpreted as a measure of the likelihood of occurrence of the event A in a experiment whose possible outcomes are the elements of A. If $P(A) = 0$, A is virtually impossible, and A is called a *P-null set*. If $P(A) = 1$, A is a certainty and it is said that A occurs *with probability* 1 or *almost surely*.

An interesting basic property that follows from the definition of P as a measure (in particular, property (1.5)) is the Borel-Cantelli lemma:

THEOREM 1.1   For an infinite sequence of events $\{A_n\} \subseteq \mathcal{A}$,

$$\sum_{n=1}^{\infty} P(A_n) < \infty \Longrightarrow P\left(\bigcap_{n=1}^{\infty} \bigcup_{k=n}^{\infty} A_k\right) = 0.$$

The set $\cap_{n=1}^{\infty} \cup_{k=n}^{\infty} A_k$ represents the set of outcomes occurring in infinitely many of the events A_n and is sometimes denoted by $\limsup A_n$. The Borel-Cantelli lemma indicates, then, that the outcomes occurring infinitely often in the A_n form a P-null set if $\sum P(A_n)$ converges.

Now let $(\Omega_1, \mathcal{A}_1)$ and $(\Omega_2, \mathcal{A}_2)$ be *measurable spaces*, that is, $\mathcal{A}_i$ is a σ-algebra of subsets of Ω_i, $i = 1, 2$. If X is a mapping of Ω_1 into Ω_2, denoted by $X : \Omega_1 \to \Omega_2$, then X is *measurable* if, for each $A \in \mathcal{A}_2$,

$$\{\omega : X(\omega) \in A\} = X^{-1}(A) \in \mathcal{A}_1.$$

The collection $\mathcal{A}(X) = \{X^{-1}(A) : A \in \mathcal{A}_2\}$ is a σ-algebra in Ω_1, called the *σ-algebra generated by* X and is the smallest such σ-algebra such that X is measurable. Of particular interest will be the cases of Ω_2 = set of real numbers R, Ω_2 = set of n-dimensional vectors with real components R^n, and Ω_2 = set of $n \times m$ matrices with real entries $R^{n \times m}$. In these cases $\mathcal{A}_2$ is usually taken to be the σ-algebra of Borel sets of the appropriate dimension: the σ-algebra generated by the set of intervals (rectangles, etc.) and denoted by $\mathcal{B}$, $\mathcal{B}^n$ or $\mathcal{B}^{n \times m}$. Measurable functions $X : \Omega_1 \to \Omega_2$ are called *random variables, random vectors*, and *random matrices*, respectively, in each of these cases; X is a random vector (resp. matrix) if and only if each component X_i (resp. X_{ij}) is a random variable. A function $X : \Omega_1 \to R$ is a random variable if and only if

$$X^{-1}(-\infty, a] = \{\omega : X(\omega) \le a\} \in \mathcal{A}_1 \qquad \text{for all real } a. \qquad (1.7)$$

If X attains $\pm\infty$ and (1.7) holds for intervals of the form $[-\infty, a]$, then X is called an *extended random variable*. Extended random vectors and matrices are defined likewise.

      EXAMPLE 1.1 The *indicator function* I_A of a set $A \subseteq \Omega_1$ defined by

$$I_A(\omega) = \begin{cases} 1, & \text{if } \omega \in A \\ 0, & \text{if } \omega \notin A \end{cases}$$

is a random variable if and only if $A \in \mathcal{A}_1$, that is, if and only if A is an event, or A is measurable.

      If the measurable space $(\Omega, \mathcal{A})$ represents an experiment, a random variable X defined on $(\Omega, \mathcal{A})$ can be considered as an abstraction of information from the possible outcomes. This abstraction can be very coarse as in the case of an indicator function or a constant function. However, the measurability requirement means that only those subsets of Ω that are events can be distinguished by different values of X. This restriction indicates how discerning X can be, in that the information made available by X about the experiment cannot exceed that provided by the full σ-algebra $\mathcal{A}$ which describes all possible outcomes of the experiment.

      That the usual operations of analysis preserve the property of random variable is important in probability theory. For example, sums, differences, products, and, when they exist, quotients, extrema, and limits of random variables are random variables. This situation can be exploited to determine probabilities of compound events from the probabilities of simpler constituent events.

      Now let $(\Omega_1, \mathcal{A}_1, P)$ be a probability space and $(\Omega_2, \mathcal{A}_2)$ be a measurable space. If $X : \Omega_1 \to \Omega_2$ is measurable, the function P_X defined by

$$P_X(A) = P\{\omega : X(\omega) \in A\} = P(X \in A)$$

for each $A \in \mathcal{A}_2$ is called the *distribution* of X. P_X uniquely specifies X in that it carries all possible probabilistic information about X. If, in particular, $X = (X_1, \ldots, X_n)$ is a random vector its *distribution function*

$F_X : R^n \to [0,1]$ is defined by

$$F_X(x) = P_X \left(\prod_{i=1}^{n} (-\infty, x_i] \right) = P(X_1 \le x_1, \ldots, X_n \le x_n) \qquad (1.8)$$

where $x = (x_1, \ldots, x_n)$; F_X is sometimes called the *joint distribution* of the random variables $X_1, \ldots, X_n$. The convenience of working with distribution functions lies in the fact that they are point functions as opposed to set functions, and are defined on Euclidean space. Properties of F_X that follow from (1.8) are as follows:

F_X is nondecreasing and continuous from the right in

each x_i : for $1 \le i \le n$,

$x_i < y_i$ implies $F_X(x_1, \ldots, x_i, \ldots, x_n) \le F_X(x_1, \ldots, y_i, \ldots, x_n)$,

and $\displaystyle\lim_{y_i \downarrow x_i} F_X(x_1, \ldots, y_i, \ldots, x_n) = F_X(x_1, \ldots, x_i, \ldots, x_n)$. (1.9)

$$\lim_{x_1 \to -\infty} F_X(x) = 0, \qquad \lim_{x_1 \to +\infty} F_X(x) = 1.$$
$$\vdots \qquad\qquad \vdots$$
$$x_n \to -\infty \qquad x_n \to +\infty$$

Any function $F : R^n \to [0,1]$ satisfying these properties is the distribution of a random vector. In applications, finite sets of random variables are usually identified with their joint distribution function. The *marginal distribution* $F_{X_{n_1}, \ldots, X_{n_k}}$ may be obtained from F_X for any subset $X_{n_1}, \ldots, X_{n_k}$ of the random variables $X_1, \ldots, X_n$ as follows:

$$F_{X_{n_1}, \ldots, X_{n_k}}(x_1, \ldots, x_k)$$
$$= \lim_{\substack{y_j \to \infty \\ j \notin \{n_1, \ldots, n_k\}}} F_X(y_1, \ldots, x_1, \ldots, x_k, \ldots, y_n)$$

with arrows pointing to the n_1 location and n_k location.

X is said to be *absolutely continuous* if X has a density, that is, if there exists a function $f : R^n \to R_+ = [0, \infty]$ such that

$$F_X(x_1, \ldots, x_n) = \int_{-\infty}^{x_1} \cdots \int_{-\infty}^{x_n} f(y_1, \ldots, y_n) \, dy_n \ldots dy_1. \qquad (1.10)$$

Then at every point of continuity of F

$$\left. \frac{\partial^n F_X}{\partial x_1 \ldots \partial x_n} \right|_{x = (x_1, \ldots, x_n)} = f(x_1, \ldots, x_n)$$

Furthermore, if X is absolutely continuous, the marginal distributions are given by

$$F_{X_{n_1},\dots,X_{n_k}}(x_1,\dots,x_k) = \int_{-\infty}^{\infty} \cdots \underset{\underset{\substack{n_1 \\ \text{location}}}{\uparrow}}{\int_{-\infty}^{x_1}} \cdots \underset{\underset{\substack{n_k \\ \text{location}}}{\uparrow}}{\int_{-\infty}^{x_k}}$$

$$\cdots \int_{-\infty}^{\infty} f(y_1,\dots,y_n)\, dy_n \cdots dy_1.$$

EXAMPLE 1.2   A random vector $X = (X_1,\dots,X_n)$ has a *normal (or Gaussian) distribution* if there exist a $\mu \in R^n$ and an $n \times n$ positive definite symmetric matrix Σ such that X has the density $f(x) = [(2\pi)^n \det \Sigma]^{-1/2} \exp\left\{-\frac{1}{2}(x-\mu)^T \Sigma^{-1}(x-\mu)\right\}$. In this case X is denoted by $\mathcal{N}(\mu, \Sigma)$. (The superscript T denotes transpose here.)

1.2 EXPECTATION

Let X be a random variable defined on the probability space $(\Omega, \mathcal{A}, P)$. *The expectation of X,*

$$E(X) = \int_{\Omega} X\, dP \tag{2.1}$$

if it exists, is the average of X over the entire probability space. As such, the expectation, *expected value*, or *mean* of X constitutes the coarsest probabilistic information concerning the random variable X. The definition of expectation may be built up successively from simple random variables as follows. For indicator functions $X = I_A$, $A \in \mathcal{A}$,

$$E(X) = \int_{\Omega} X\, dP = \int_{\Omega} I_A\, dP = \int_{A} dP = P(A).$$

Finite linear combinations of indicator functions,

$$X = \sum_{i=1}^{n} c_i I_{A_i}$$

$c_i \in R$, $A_i \in \mathcal{A}$, represent random variables that attain finitely many values, and are called *simple functions*. In this case,

$$E(X) = \int_{\Omega} \sum c_i I_{A_i}\, dP = \sum c_i \int_{\Omega} I_{A_i}\, dP = \sum c_i P(A_i).$$

If X is a nonnegative random variable,

$$E(X) = \sup_{\substack{Y \leq X \\ Y \text{ simple}}} E(Y)$$

Then, since any random variable X can be written as the difference of nonnegative random variables,

$$X = X^+ - X^-$$

where $X^+ = XI_{\{X \geq 0\}}$ and $X^- = -XI_{\{X < 0\}}$, one can define

$$E(X) = E(X^+) - E(X^-)$$

provided at least one of the expected values on the right is finite. The expectation $E(X)$ exists (is finite) if and only if $E(|X|) < \infty$, a situation described by the notation

$$X \in L^1(\Omega, \mathcal{A}, P),$$

or just $X \in L^1$; in this case,

$$|E(X)| \leq E(|X|). \tag{2.2}$$

Some properties of expectation are as follows:

If a and b are constants, $E(aX + bY) \leq aE(X) + bE(Y).$ (2.3)

If $X \leq Y, E(X) \leq E(Y).$ (2.4)

If $g(x)$ is a convex function, that is, if for $0 < \lambda < 1$,

$$g(\lambda x + (1 - \lambda)y) \leq \lambda g(x) + (1 - \lambda)g(y),$$

then $g(E(X)) \leq E(g(X))$ (Jensen's inequality). (2.5)

Even more fundamental is Markov's inequality:

If $X \geq 0$ and $a > 0, P(X \geq a) \leq E(X)/a.$ (2.6)

If X is a random vector or matrix $E(X)$ is defined componentwise.

A change of variables formula is useful in the actual computation of expectations. Let X be a random n-vector, and let the random m-vector Y be given by $Y = g(X)$, where $g : R^n \to R^m$ is measurable. Then

$$E(Y) = \int_\Omega g(X)\, dP = \int_{R^n} g(x)\, dF_X(x), \tag{2.7}$$

where the integral on the right side is the Lebesgue-Stieltjes integral, in general. (If it exists, the corresponding Riemann-Stieltjes integral is identical.)

In particular, taking g to be the identity function

$$E(X) = \int_{R^n} x\, dF_X(x);$$

and if X has the density f, from (1.10) and (2.7),

$$E(X) = \int_{R^n} xf(x)\, dx.$$

An important special case occurs when $g(x) = |x|^p$, $p \geq 1$. The corresponding collection of random variables X for which $Y = g(X)$ has finite expectation forms a complete normed linear (Banach) space, and is denoted by $L^p(\Omega, \mathcal{A}, P)$. The norm in this space is

$$\|X\|_p = \{E(|X|^p)\}^{1/p},$$

and the triangle inequality, here,

$$\|X + Y\|_p \leq \|X\|_p + \|Y\|_p, \tag{2.8}$$

is called *Minkowski's inequality*. If $p \geq q$, then $L^p \subseteq L^q$. If X and Y are random n-vectors in L^p and L^q respectively, where $p > 1$ and $1/p + 1/q = 1$, then

$$E(X^T Y) = \sum_{i=1}^{n} E(X_i Y_i) \qquad \text{exists and is finite,}$$

and satisfies *Hölder's inequality*

$$|E(X^T Y)| \leq \|X\|_p \|Y\|_q. \tag{2.9}$$

In the case of $p = 2$, $E(X^T Y)$ defines an inner product, (2.9) is known as the Cauchy-Schwarz inequality, and the space L^2 is a Hilbert space. In the probability context, the L^2 norm is often referred to as the *quadratic mean* or *mean square* norm. If X and Y are in L^2, then the *covariance* of X and Y is an $n \times n$ matrix defined by

$$\text{Cov}(X, Y) = E([X - E(X)][Y - E(Y)]^T). \tag{2.10}$$

When X and Y are random variables, (2.10) can be written

$$\text{Cov}(X, Y) = E(XY) - E(X)E(Y). \tag{2.11}$$

If $\text{Cov}(X, Y) = 0$, X and Y are said to be *uncorrelated*. $\text{Cov}(X, X)$ is written $\text{Cov}(X)$, and for the random variable case is known as the *variance* $V(X)$; from (2.11)

$$V(X) = E[X - E(X)]^2 = E(X^2) - [E(X)]^2. \tag{2.12}$$

The square root, $\sigma = (V(X))^{1/2}$ is called the *standard deviation* of X. Suppose $\mu = E(X)$ and $\sigma^2 = V(X)$ are both finite. Applying Markov's inequality (2.6) to the nonnegative random variable $(X - \mu)^2$ yields Chebyshev's inequality:

$$\text{if } a > 0, P((X - \mu)^2 \geq a) \leq \sigma^2/a. \tag{2.13}$$

If $n = 1$ and if k is a natural number and $X \in L^k$, then $E(X^k)$ and $E([X - E(X)]^k)$ define the kth *moment* and k^{th} *central moment* of X, respectively.

The *characteristic function* of a random n-vector X with distribution finite F_X is given by

$$\varphi_X(u) = E(\exp\{iu^T X\}) = \int_{R^n} \exp\{iu^T x\}\, dF_X(x),$$

$$u \in R^n \tag{2.14}$$

and the distribution is uniquely determined by φ.

EXAMPLE 1.3 The random vector X with normal distribution $N(\mu, \Sigma)$ as defined in Example 1.2 has expectation $E(X) = \mu$, covariance matrix $\text{Cov}(X) = \Sigma$, and characteristic function

$$\varphi_X(u) = \exp\{i(u^T\mu - \tfrac{1}{2}u^T\Sigma u)\}.$$

In case $n = 1$, X with normal distribution $N(\mu, \sigma^2)$ has central moments,

$$E([X - \mu]^k) = \begin{cases} 0, & k \text{ odd} \\ 1 \cdot 3 \cdot 5 \cdots (n-1)\sigma^k, & k \text{ even.} \end{cases} \tag{2.15}$$

1.3 INDEPENDENCE, CONDITIONAL EXPECTATION, AND CONDITIONAL PROBABILITY

The concept of independence of events and random variables is important from the point of view of computing probabilities of compound events from simpler constituent events. The events $A_1, \ldots, A_n$ in the probability space $(\Omega, \mathcal{A}, P)$ are said to be *independent* if for every subset $\{i_1, i_2, \ldots, i_k\}$ of the set of integers $\{1, 2, \ldots, n\}$

$$P(A_{i_1} \cap \cdots \cap A_{i_k}) = P(A_{i_1}) \cdots P(A_{i_k}). \tag{3.1}$$

If the events $A_1, \ldots, A_n$ are sequential in time, (3.1) implies that the likelihood of any event occurring is not dependent on specific outcomes of the first few or any subset of other events, for that matter. The sub-σ-algebras

$A_1, \ldots, A_n$ of A are said to be independent if (3.1) holds for any choice of events $A_i \in A_i$, $i = 1, \ldots, n$. The random vectors $X_1, \ldots, X_n$ defined on (Ω, A, P) are independent if the corresponding σ-algebras $A(X_1), \ldots, A(X_n)$ generated by the random vectors are independent. (The random vectors may have different dimensions.) If F and $F_1, \ldots, F_n$ denote the joint and marginal distributions of the random variables $X_1, \ldots, X_n$ respectively, then $X_1, \ldots, X_n$ are independent if and only if

$$F(x_1, \ldots, x_n) = F_1(x_1) \cdots F_n(x_n). \tag{3.2}$$

If the joint density f exists, and $f_1, \ldots, f_n$ denote the marginal densities, independence of $X_1, \ldots, X_n$ is equivalent to (3.2) with the corresponding F's replaced by these functions. If $X_1, \ldots, X_n$ are independent random variables

$$E\left(\prod_{i=1}^{n} X_i\right) = \prod_{i=1}^{n} E(X_i). \tag{3.3}$$

Furthermore, if these random variables are in L^2, they are uncorrelated and

$$V\left(\sum_{i=1}^{n} X_i\right) = \sum_{i=1}^{n} V(X_i). \tag{3.4}$$

Arbitrary collections of events $\{A_\alpha\}$ or random vectors $\{X_\alpha\}$ are independent if all finite subcollections are independent. The statements made in this section as well as subsequent sections unless otherwise indicated hold for random matrices also.

In the probability space (Ω, A, P), let $A \in A$, $B \in A$, and $P(B) > 0$. When the events A and B are not independent, the *conditional probability of A given B*

$$P(A \mid B) = \frac{P(A \cap B)}{P(B)} \tag{3.5}$$

is of interest. In place of (3.1) one has generally, then,

$$
\begin{aligned}
P(A_1 \cap \cdots \cap A_n) &= P(A_1 \cap \cdots \cap A_n \mid A_1 \cap \cdots \cap A_{n-1}) \\
&\quad \times P(A_1 \cap \cdots \cap A_{n-1}) \\
&= P(A_1 \cap \cdots \cap A_n \mid A_1 \cap \cdots \cap A_{n-1}) \\
&\quad \times P(A_1 \cap \cdots \cap A_{n-1} \mid A_1 \cap \cdots \cap A_{n-2}) \\
&\quad \cdots P(A_1 \cap A_2 \mid A_1) P(A_1),
\end{aligned}
\tag{3.6}
$$

for any collection of events $\{A_i\}$ for which the conditional probabilities are defined. If the sequence $\{B_n\} \subseteq \mathcal{A}$ partitions Ω, that is,

$$\cup B_n = \Omega \qquad \text{and} \qquad B_i \cap B_j = \varnothing \qquad \text{if } i \neq j,$$

and if $P(B_n) > 0$, each n, the probability of A can be expressed in terms of the conditional probabilities $P(A \mid B_n)$:

$$P(A) = \sum_n P(A \mid B_n) P(B_n);$$

this is called the law of total probability.

To deal with more complicated conditioning of random variables and σ-algebras, the more general concept of conditional expectation is introduced. Let X be a random vector in $L^1(\Omega, \mathcal{A}, P)$, and let $\mathcal{B}$ be a sub-σ-algebra of $\mathcal{A}$. Note that X is not necessarily a random vector on the probability space $(\Omega, \mathcal{B}, P)$, as it is not necessarily measurable with respect to $\mathcal{B}$. Now denote by $E(X \mid \mathcal{B})$ that random vector Y on the space $(\Omega, \mathcal{B}, P)$ satisfying

$$\int_B Y \, dP = \int_B X \, dP, \qquad B \in \mathcal{B}. \tag{3.7}$$

This random vector $Y = E(X \mid \mathcal{B})$ is called the *conditional expectation of X given $\mathcal{B}$*. [The existence of Y is guaranteed by the Radon-Nikodym theorem (Ash, 1972, p. 63) of real analysis.] As X represents a portion of all information in the events in $\mathcal{A}$, $E(X \mid \mathcal{B})$ constitutes that portion of the information carried by X which is related to the sub-σ-algebra $\mathcal{B}$. How coarse this information is depends on how X and $\mathcal{B}$ are related. At one extreme, if X is $\mathcal{B}$-measurable, then

$$E(X \mid \mathcal{B}) = X; \tag{3.8}$$

at the other, if X and $\mathcal{B}$ are independent

$$E(X \mid \mathcal{B}) = E(X). \tag{3.9}$$

Generally, for sub-σ-algebras $\mathcal{B}_1 \subseteq \mathcal{B}_2 \subseteq \mathcal{A}$,

$$E(E(X \mid \mathcal{B}_1) \mid \mathcal{B}_2) = E(E(X \mid \mathcal{B}_2) \mid \mathcal{B}_1) = E(X \mid \mathcal{B}_1). \tag{3.10}$$

It is clear that

$$E(E(X \mid \mathcal{B})) = E(X). \tag{3.11}$$

Other properties are inherited by the conditional expectation due to the integral form of the definition (3.7). For completeness, some of these are listed here:

For constants a and b, $E(aX + bY \mid \mathcal{B})$

$$= aE(X \mid \mathcal{B}) + bE(Y \mid \mathcal{B}). \tag{3.12}$$

If $X \leq Y$, then $E(X \mid \mathcal{B}) \leq E(Y \mid \mathcal{B})$. $\hspace{2.5cm}$ (3.13)

If g is convex, $g(E(X \mid \mathcal{B})) \leq E(g(X) \mid \mathcal{B})$

(Jensen's inequality); $\hspace{4.5cm}$ (3.14)

in particular,

$$|E(X \mid \mathcal{B})| \leq E(|X| \mid \mathcal{B}). \tag{3.15}$$

(Properties (3.12) to (3.15) hold with probability 1, or almost surely.)

The *conditional probability* $P(A \mid \mathcal{B})$ *of an event A given the sub-σ-algebra* $\mathcal{B} \subseteq \mathcal{A}$ is defined by

$$P(A \mid \mathcal{B}) = E(I_A \mid \mathcal{B}). \tag{3.16}$$

The following properties are satisfied almost surely:

$P(A \mid \mathcal{B}) \geq 0$

$$P\left(\bigcup_{n=1}^{\infty} A_n \mid \mathcal{B} \right) = \sum_{n=1}^{\infty} P(A_n \mid \mathcal{B}), \qquad \text{if } A_i \cap A_j \neq \varnothing, i \neq j \tag{3.17}$$

$P(\Omega \mid \mathcal{B}) = 1.$

EXAMPLE 1.4   Let $B \in \mathcal{A}$ with $0 < P(B) < 1$, and let $\mathcal{B} = \{\varnothing, B, B^c, \Omega\}$;

$$P(A \mid \mathcal{B})(\omega) = E(I_A \mid \mathcal{B})(\omega) = \begin{cases} \dfrac{P(A \cap B)}{P(B)}, & \omega \in B \\[2mm] \dfrac{P(A \cap B^c)}{P(B^c)}, & \omega \in B^c. \end{cases}$$

Because the preceding properties (3.17) hold only with probability 1, $P(\cdot \mid \mathcal{B})(\omega)$ is not a probability, for fixed ω, generally.

Now let $Y : (\Omega, \mathcal{A}) \rightarrow (R^m, \mathcal{B}^m)$ be a random vector with distribution P_Y. The *conditional probability of an event $A \in \mathcal{A}$ given $Y = y$*, $P(A \mid Y = y)$, and the *conditional expectation of a random vector X given $Y = y$*, $E(X \mid Y = y)$, may be defined for $y \in R^m$. Similarly to the definition of $E(X \mid \mathcal{B})$, if $E(X)$ exists, there is a unique (with P_Y probability 1) function

$$g : (R^m, \mathcal{B}^m) \longrightarrow (R^m, \mathcal{B}^m)$$

such that

$$\int_{\{Y^{-1}(B)\}} X \, dP = \int_B g(y) \, dP_Y(y), \qquad B \in \mathcal{B}^m; \tag{3.18}$$

define $E(X \mid Y = y) = g(y)$. In particular,

$$E(X) = \int_\Omega X \, dP = \int_{R^m} E(X \mid Y = y) \, dP_Y(y).$$

If $A \in \mathcal{A}$, one can then define the conditional probability

$$P(A \mid Y = y) = E(I_A \mid Y = y). \tag{3.19}$$

EXAMPLE 1.5   Suppose Y is a discrete random variable: Y takes on countably many values $y_1, y_2, \ldots$ and $P(Y = y_i) > 0$, each i. Then, for any $A \in \mathcal{A}$

$$P(A \mid Y = y_i) = \frac{P(A \cap \{Y = y_i\})}{P(Y = y_i)}.$$

If, in addition, X also is a discrete random variable

$$E(X \mid Y = y_i) = \sum_j x_j \frac{P(X = x_j, Y = y_i)}{P(Y = y_i)}$$

$$= \sum_j x_j P(X = x_j \mid Y = y_i).$$

If X and Y have joint density $f(x, y)$, and $f_2(y)$ represents the marginal density of Y

$$f_2(y) = \int_{R^n} f(x, y) \, dx,$$

then the *conditional density* $f(x \mid y)$ *of* X *given* $Y = y$ is defined

$$f(x \mid y) = \frac{f(x, y)}{f_2(y)}, \qquad f_2(y) \neq 0, \tag{3.20}$$

and satisfies

$$E(k(X) \mid Y = y) = \int_{R^n} k(x) f(x \mid y) \, dx$$

for any integrable function k.

As a final comment on this section, if $g(y) = E(X \mid Y = y)$ and $h(\omega) = g(Y(\omega))$, then

$$h = E(X \mid Y) = E(X \mid \mathcal{B})$$

where $\mathcal{B} = \mathcal{A}(Y)$, which ties together the three types of conditional expectation.

EXAMPLE 1.6   Suppose X and Y are bivariate normally distributed random variables: their joint density is given by

$$f(x, y) = \left(2\pi\sigma_X\sigma_Y \sqrt{1 - \rho^2}\right)^{-1} \exp\left\{-\frac{1}{2(1 - \rho^2)}\right.$$

$$\times \left.\left[\left(\frac{x - \mu_X}{\sigma_X}\right)^2 + \left(\frac{y - \mu_Y}{\sigma_Y}\right)^2 - 2\rho\frac{(x - \mu_X)(y - \mu_Y)}{\sigma_X\sigma_Y}\right]\right\}.$$

The conditional distribution of X given $Y = y$ is normal with mean $\mu_X + \rho(\sigma_X/\sigma_Y)(y - \mu_Y)$ and variance $\sigma_X^2(1 - \rho^2)$, that is,

$$f(x \mid y) = (2\pi\sigma_X^2(1 - \rho^2))^{-1/2}$$

$$\times \exp\left\{-\frac{1}{2(1 - \rho^2)}\left[\frac{x - \mu_X - \rho(\sigma_X/\sigma_Y)(y - \mu_Y)}{\sigma_X}\right]^2\right\}.$$

The conditional expectation of X given $Y = y$ is

$$E(X \mid Y = y) = \mu_X + \rho\frac{\sigma_X}{\sigma_Y}(y - \mu_Y),$$

and the conditional expectation of X given Y is the random variable

$$E(X \mid Y) = \mu_X + \rho\frac{\sigma_X}{\sigma_Y}(Y - \mu_Y).$$

1.4   LIMIT THEOREMS

It has been mentioned that solutions of stochastic differential equations may be regarded as indexed sets of random variables with the index or parameter set corresponding to some interval. Some important questions here are concerned with the asymptotic behavior of these random variables as the parameter approaches a specific value or tends to infinity. These random variables, each of which represents the solution at some time instant, are clearly not independent, and the index set, being an interval, is continuous as opposed to discrete. Nevertheless, the classical limit theorems of probability theory, which deal with the asymptotic behavior of sequences of independent identically distributed random variables (i.i.d.'s) provide a useful frame of reference for analogous qualitative results for stochastic differential equations to be given later. Here the most important of these limit theorems are listed. But first a brief discussion of the major modes of convergence for sequences of random variables is in order.

There are three such modes of convergence. Suppose $\{X_n\}$ is a sequence of random variables defined on a common probability space $(\Omega, \mathcal{A}, P)$. *With probability 1*, or *almost sure convergence* of X_n to a random variable X defined on $(\Omega, \mathcal{A}, P)$ means

$$P\left(\omega : |X_n(\omega) - X(\omega)| \longrightarrow 0 \text{ as } n \longrightarrow \infty\right) = 1. \tag{4.1}$$

L^2 or *mean square convergence* of X_n to X means

$$\|X_n - X\|_2 \equiv \left[E \mid X_n - X|^2\right]^{1/2} \longrightarrow 0, \qquad \text{as } n \longrightarrow \infty. \tag{4.2}$$

Finally, X_n *converges to X in probability* or X_n *stochastically converges to X* if, for any $\varepsilon > 0$,

$$P(\omega : |X_n(\omega) - X(\omega)| \geq \varepsilon) \longrightarrow 0, \qquad \text{as } n \longrightarrow \infty. \tag{4.3}$$

It is not difficult to see that (4.3) is equivalent to

$$E\left(\frac{|\dot{X}_n - X|}{1 + |X_n - X|}\right) \longrightarrow 0, \qquad \text{as } n \longrightarrow \infty; \tag{4.4}$$

the expression on the left in (4.4) defines a metric or a distance function between random variables. Stochastic convergence is the weakest of the three modes of convergence cited above: one has the diagram

(4.1)   (4.2)

$$\searrow \qquad \nearrow \tag{4.5a}$$

(4.3);

for the case of dominated convergence, that is, when $|X_n| \leq Y$ for some $Y \in L^2(\Omega, \mathcal{A}, P)$, (4.1) is the strongest mode of convergence and the other two are equivalent:

(4.1) $\Longrightarrow$ (4.2)

$$\searrow \qquad \nearrow \tag{4.5b}$$

(4.3);

the implication (4.1) $\Rightarrow$ (4.2) under this condition is known as the dominated convergence theorem. Furthermore, if $\{X_n\}$ satisfies either (4.2) or (4.3), there exists a subsequence $\{X_{n_k}\}$ which satisfies (4.1).

The dominated convergence theorem holds also in L^1, and more generally, along with the two other basic convergence theorems of measure theory, Fatou's lemma and the monotone convergence theorem, it holds for conditional expectations as well. These results are now listed.

Let X_n, X and Y be random variables on $(\Omega, \mathcal{A}, P)$,

and suppose $\mathcal{B}$ is a sub-σ-algebra of $\mathcal{A}$:

If $|X_n| \leq Y$, for all n, where $E(Y) < \infty$, and $X_n \to X$ w.p. 1, then

$$E(X_n \mid \mathcal{B}) \longrightarrow E(X \mid \mathcal{B}) \qquad \text{w.p. 1}$$

(dominated convergence theorem). $\qquad\qquad$ (4.6)

If $X_n \geq Y$, for all n, where $E(Y) > -\infty$, then

$$\liminf_{n \to \infty} E(X_n \mid \mathcal{B}) \geq E(\liminf_{n \to \infty} X_n \mid \mathcal{B}) \qquad \text{w.p. 1}$$

(Fatou's lemma). $\qquad\qquad$ (4.7)

If, further, $X_n \uparrow X$ w.p.1, then

$$E(X_n \mid \mathcal{B}) \uparrow E(X \mid \mathcal{B}) \qquad \text{w.p. 1}$$

(monotone convergence theorem). $\qquad\qquad$ (4.8)

(In the above, $X_n \uparrow X$ denotes $X_n \leq X_{n+1}$, all n, and $\text{limit}_{n \to \infty} X_n = X$.)

It is emphasized that the convergence results given in the last two paragraphs are fundamentally results of measure theory, as opposed to probability theory. Only the assumption that a probability is a finite measure was used. In particular, all these results are valid no matter what other probabilistic relationships exist among the random variables X_n, that is, no matter what the form of the corresponding joint distributions. The classical limit theorems of probability theory, on the other hand, assume specific probabilistic relations among the X_n and give more particular probabilistic information about the limit. These limit theorems are given next.

Let X_n be a sequence of i.i.d.'s, each defined on the same probability space $(\Omega, \mathcal{A}, P)$. Let

$$S_n = X_1 + \cdots + X_n. \qquad\qquad (4.9)$$

The classical limit theorems of probability describe the asymptotic behavior of the sums S_n suitably normalized. The most basic one is

THEOREM 1.2   (Strong Law of Large Numbers)   Let $E(X_i) = \mu$. Under the hypothesis (4.9),

$$\lim_{n \to \infty} \frac{S_n}{n} = \mu \qquad \text{w.p. 1.} \tag{4.10}$$

This theorem indicates that the two notions of sample average and expected value coincide asymptotically; that is, for sufficiently large n, the sample average of $X_1, \ldots, X_n$ approximates arbitrarily closely the common expected value $E(X_i)$:

$$E(X_i) \approx \frac{X_1 + \cdots + X_n}{n}. \tag{4.11}$$

The importance of this fact stems from the observation that one can calculate sample averages.

EXAMPLE 1.7   Suppose the X_i are independent Bernoulli random variables with parameter p; $P(X_i = 1) = p$ and $P(X_i = 0) = 1-p$. The X_i's can be considered as mathematically modeling the outcomes $(1 = \text{success}$ and $0 = \text{failure})$ of repeated independent trials of an experiment, where the probability of success is p. Now

$$E(X_i) = p, \qquad \text{each } i,$$

and if n_s denotes the number of successes in the first n trials,

$$n_s = X_1 + \cdots + X_n = S_n.$$

So (4.11) becomes, in this case,

$$p \approx \frac{n_s}{n} \tag{4.12}$$

for sufficiently large n. The right side of (4.12) is the frequency "definition" of probability. The idea that (4.12) should define probability, rather than be the consequence of some other definition, presented an obstacle to the development of probability theory for some time.

Suppose now that

$$\mu = E(X_i) < \infty \qquad \text{and} \qquad \sigma = (V(X_i))^{1/2} < \infty.$$

Then the sequence $\{Y_n\}$ given by

$$Y_n = \frac{S_n - n\mu}{\sqrt{n}\sigma}$$

is normalized so that each Y_n has mean zero and variance one. In particular the Y_n do not degenerate, in contrast to S_n/n, as $n \to \infty$. The next of these remarkable results indicates that the Y_n do not grow faster than $\sqrt{2 \log\log n}$, regardless of the distribution of X_n.

THEOREM 1.3   (Law of the Iterated Logarithm)   Under the preceding hypotheses on $\{Y_n\}$,

$$\limsup_{n \to \infty} \frac{Y_n}{\sqrt{2 \log\log n}} = +1 \qquad \text{w.p. 1,}$$

and   (4.13)

$$\liminf_{n \to \infty} \frac{Y_n}{\sqrt{2 \log\log n}} = -1 \qquad \text{w.p. 1.}$$

The most important and well known of the classical limit theorems of probability theory is the following.

THEOREM 1.4   (Central Limit Theorem)   Under the preceding hypotheses on $\{Y_n\}$,

$$\lim_{n \to \infty} P(Y_n \leq y) = \frac{1}{\sqrt{2\pi}} \int_{-\infty}^{y} e^{-t^2/2} \, dt, \qquad y \in R. \tag{4.14}$$

This result states that the sum of a large number of zero mean, small variance i.i.d. random variables is stochastically approximately standard Gaussian or normal, no matter what the common distribution of those random variables.

EXAMPLE 1.8   Let X_i be the independent Bernoulli random variables with distributions:

$$P(X_i = 1) = \tfrac{1}{2} = P(X_i = -1). \tag{4.15}$$

Then $\mu = E(X_i) = 0$, and $\sigma = (E(X_i^2))^{1/2} = 1$, so that

$$Y_n = \frac{S_n - n\mu}{\sqrt{n}\sigma} = \frac{S_n}{\sqrt{n}}.$$

In this case S_n is called a *random walk*, and the central limit theorem indicates that the random walk is stochasticaly asymptotic to $\mathcal{N}(0, n)$ random variables.

Sequence of i.i.d.'s are stochastic processes in a trivial sense. The specification of how individual random variables of a nontrivial porcess, such as a solution of a stochastic differential equation, are related is the main goal of the theory of stochastic processes, to be briefly discussed next.

1.5   BASIC PROPERTIES OF STOCHASTIC PROCESSES

Suppose $(\Omega, \mathcal{A}, P)$ is a probability space, and J is an arbitrary set. A family

$$X(t, \omega), \quad \omega \in \Omega, \quad t \in J$$

of random variables or random n-vectors is called a *stochastic process* with index or parameter set J and state space R or R^n. In this book the parameter set J is an interval. Stochastic processes are functions of two variables; the usual notation suppresses the probability space variable ω. For each fixed $t \in J$,

$$X(t) = X(t, \cdot)$$

denotes a random variable or random n-vector on the probability space $(\Omega, \mathcal{A}, P)$; for each fixed $\omega \in \Omega$,

$$X(\cdot, \omega)$$

corresponds to a real-valued or n-vector-valued function defined on J. The latter is called a *sample path*, *trajectory*, or *realization* of the process. The theory of stochastic processes relates these inherent random and sample path structures. Generally, as briefly mentioned in the last section, the individual random variables or vectors of the process are dependent. Understanding the stochastic process requires, therefore, knowing the (uncountably many, since J is an interval) joint distributions of these random variables or vectors. The collection of all such joint distributions constitutes the *probability law* of the process. Conversely, one may ask when a given collection of distribution functions uniquely (in the probability law sense) determines a stochastic process. The existence or consistency theorem of Kolmogorov, to be given next, answers this question.

For simplicity of notation only, suppose throughout the rest of this section that $X(t)$ is a scalar process (the state space is R) on J. The family of finite joint (finite-dimensional) distributions

$$\left\{ F_{t_1, \ldots, t_n} : t_1, \ldots, t_n \in I, t_i \neq t_j, i \neq j \right\}$$

is defined, as in (1.8), by

$$F_{t_1, \ldots, t_n}(x_1, \ldots, x_n) = F_{X(t_1), \ldots, X(t_n)}(x_1, \ldots, x_n)$$
$$= P(X(t_1) \leq x_1, \ldots, X(t_n) \leq x_n).$$

In addition to individually satisfying properties (1.9), the finite dimensional distributions collectively satisfy the *symmetry property*

$$F_{t_{i_1}, \ldots, t_{i_n}}(x_{i_1}, \ldots, x_{i_n}) = F_{t_1, \ldots, t_n}(x_1, \ldots, x_n) \tag{5.1}$$

for any permutation $\{i_1, \ldots, i_n\}$ of the integers $1, \ldots, n$, and the *compatability property*

$$\lim_{\substack{x_k \to \infty \\ m+1 \leq k \leq n}} F_{t_1, \ldots, t_n}(x_1, \ldots, x_n) = F_{t_1, \ldots, t_m}(x_1, \ldots, x_m). \tag{5.2}$$

THEOREM 1.5   (Kolmogorov Extension Theorem)   Suppose $\mathcal{F}$ is a collection of distribution functions $F_{t_1, \ldots, t_n}$ [each such F satisfies (1.9)], one for each distinct finite subset $t_1, \ldots, t_n$ of J. If, in addition, the members of J satisfy the symmetry and compatibility conditions, (5.1) and (5.2), then there is a stochastic process whose family of finite-dimensional distributions coincide with J. Furthermore, the process is unique in the probability law sense.

Theorem 1.5, for example, helps to characterize two notions of "equality" for stochastic processes. Two processes $X(t, \omega)$ and $Y(t, \omega)$ are called *equivalent* provided

$$X(t, \cdot) = Y(t, \cdot) \text{ w.p. } 1, \qquad \text{each } t \in J; \tag{5.3}$$

that is, for each $t \in J$, the set

$$A_t = \{\omega : X(t, \omega) \neq Y(t, \omega)\}$$

is a P-null set, $P(A_t) = 0$. In this case each process is referred to as a *version* of the other. It is clear that equivalent processes possess the same finite joint distributions, and hence by Theorem 1.5, they share the same probability law. The second notion of "equality" of processes requires that the sample paths of the processes be identical with probability 1, that is, the set

$$A = \{\omega : X(t, \omega) \neq Y(t, \omega), \text{ some } t \in J\}$$

is a P-null set. Since for each $t \in J$, clearly

$$A = \bigcup_{t \in J} A_t$$

two processes having identical sample paths with probability 1 must be equivalent; since the A_t may be distinct and since J is uncountable, $P(A_t) = 0$, (each t) does not imply $P(A) = 0$, that is, equivalent processes need not have identical sample paths with probability 1. If, however, the processes are *continuous*, that is, the sample paths are continuous functions of t with probability 1, then the two notions of equality, equivalence and identical sample paths w.p. 1, coincide.

Corresponding to the other modes of convergence for sequences of random variables, given in Section 1.4, there are other types of continuity discussed for stochastic processes $X(t, \omega)$. For t, $t + h \in J$,

$$P(\omega : |X(t + h, \omega) - X(t, \omega)| \geq \varepsilon) \longrightarrow 0,$$

$$\text{as } h \longrightarrow 0, \text{ for any } \varepsilon > 0 \quad (5.4)$$

defines *continuity in probability* or *stochastic continuity* of X at t, and

$$\|X(t + h) - X(t)\|_2^2 = E|X(t + h, \omega) - X(t, \omega)|^2 \longrightarrow 0,$$

$$\text{as } h \longrightarrow 0 \quad (5.5)$$

constitutes *mean square* or *quadratic mean continuity of X at t*. The process X is called *stochastic continuous* (on J) or *mean square continuous* (on J) if (5.4) or (5.5) hold respectively for all $t \in J$. Relationships among the modes of convergence given in Section 1.2 imply similar ones among the corresponding continuity types. In particular, stochastic continuity is the weakest. A process $X(t, \omega)$ which is stochastically continuous on an interval J is *uniformly stochastically continuous* on compact subintervals K of J; that is, for every $\varepsilon > 0$ and $\delta > 0$, there is an $h_0 > 0$ such that

$$P(\omega : |X(t + h, \omega) - X(t, \omega)| \geq \varepsilon) < \delta$$

$$\text{whenever } t, t + h \in K \text{ and } |h| < h_0. \quad (5.6)$$

Under a slightly stronger variation of (5.6), Loéve proved that the process X has a continuous version; this result can be used to establish the following useful criterion of Kolmogorov.

THEOREM 1.6   Suppose X is a stochastic process on the probability space $(\Omega, \mathcal{A}, P)$ with index set $J = [t_0, T]$. If there are positive constants α, β, C, and h_0 such that

$$E|X(t + h, \omega) - X(t, \omega)|^\alpha \leq C|h|^{1+\beta}$$

$$\text{whenever } t, t + h \in J \text{ and } |h| < h_0, \quad (5.7)$$

then there exists a continuous process Y on $(\Omega, \mathcal{A}, P)$ with index set J that is equivalent to X; that is, X has a continuous version.

That a stochastic process on an interval constitutes an uncountable family of random variables led to the fact that equivalent processes need not have the same sample paths. This situation also implies, more generally, that sets associated with any uncountable subset of the random variables may not be measurable. For example, given $\gamma \in R$,

$$\{\omega \colon X(t,\omega) \leq \gamma, \text{ all } t \in J\} = \bigcap_{t \in J} \{\omega \colon X(t,\omega) \leq \gamma\} \qquad (5.8)$$

may not be $\mathcal{A}$-measurable, and so could not be assigned a probability. From the applications point of view it is important to be able to associate probabilities with sets such as (5.8). The concept of separability introduced by Doob provides a solution to the dilemma. Suppose there exists a countable dense subset S of J such that, for every open interval J_o and closed interval J_c,

$$\{\omega \colon X(t,\omega) \in J_c, \text{ all } t \in J \cap J_o\} = \{\omega \colon X(t,\omega) \in J_c, \text{ all } t \in S \cap J_o\}.$$
$$(5.9)$$

Then the stochastic process is called *separable*. Notice that if X is separable, taking $J_c = (-\infty, \gamma]$ and $J_o = R$ in (5.9), the set (5.8) can be written

$$\bigcap_{t \in S} \{\omega \colon X(t,\omega) \leq \gamma\};$$

this set is $\mathcal{A}$-measurable since there are only countably many sets in the intersection. Also it is not difficult to see that separable equivalent processes have the same sample paths with probability 1. An important result indicates that any (scalar) stochastic process X has an equivalent extended real-valued process Y which is separable.

Another basic difficulty that arises is that a stochastic process $X(t,\omega)$ on a probability space $(\Omega, \mathcal{A}, P)$ may not be product measurable, although it is measurable in each variable separately. This situation would prevent, for example, application of Fubini's theorem to change the order of integration as in

$$E \int_J X(t,\omega) \, dt = \int_J EX(t,\omega) \, dt.$$

Let $\mathcal{B}_J$ and $\mathcal{B}$ represent the σ-algebras of Borel subsets of J and R respectively, and denote by $\mathcal{A} \times \mathcal{B}_J$ the σ-algebra generated by all Cartesian products $A \times B_J$, for $A \in \mathcal{A}$ and $B \in \mathcal{B}_J$. The problem mentioned above is that

$$X^{-1}(B) \in \mathcal{A} \times \mathcal{B}_J \qquad (5.10)$$

may not be valid for some $B \in \mathcal{B}$. If (5.10) holds for all $B \in \mathcal{B}$, on the other hand, the process X is called *measurable* with respect to the product $P \times \mu$ of the probability measure P on $(\Omega, \mathcal{A})$ and Lebesque measure μ on $(J, \mathcal{B}_J)$. For example, suppose J is a unit interval, so that μ is a probability measure. Then X is measurable if and only if X is a random variable on the product probability space $(\Omega \times J, \mathcal{A} \times \mathcal{B}_J, P \times \mu)$. The next theorem concludes this section by relating continuity, separability and measurability properties of stochastic processes.

THEOREM 1.7 Let X be a stochastic process on the probability space $(\Omega, \mathcal{A}, P)$ with parameter set the interval J. If X is continuous, then X is measurable and X is separable with respect to every countable dense subset S of J.

1.6 STATIONARY PROCESSES WITH INDEPENDENT INCREMENTS

Determining the complete probability law for an arbitrary stochastic process is, in general, a hopeless task. If the process has some "regularity" properties, however, the magnitude of the problem may be diminished. In many applications this is the case, fortunately. The remainder of this chapter is devoted to some important classes of stochastic processes, and this section, in particular, introduces an elementary but powerful such classification.

A stochastic process $X(t)$ is said to be *strictly stationary* if its finite-dimensional distributions are invariant under time displacements; that is, for any $t \in R$ such that t_j, $t_j + t$ are in J, all j,

$$F_{t_1+t,\ldots,t_n+t}(x_1,\ldots,x_n) = F_{t_1,\ldots,t_n}(x_1,\ldots,x_n). \tag{6.1}$$

If, in addition, for each t, $X(t) \in L^2$ (X is said to be an L^2 *process*), then for each s, t

$$E(X(t)) = \mu \tag{6.2}$$

and

$$\mathrm{Cov}(X(t), X(s)) = C(t - s), \tag{6.3}$$

that is, the mean is constant and covariance is a function of $t - s$. An L^2 process satisfying the last two properties is called *wide-sense stationary*.

Another type of stochastic process whose complete probability law is specified by reduced information about the process is a process with *independent increments* (or an *additive process*); this means that, for any finite sequence $\{t_i\} \subseteq J$ with $t_i < t_{i+1}$, the differences $X(t_{i+1}) - X(t_i)$ are independent. For such a process the characteristic function of the joint distribution of the random vectors $X(t_1), \ldots, X(t_n)$ can be written

$$\varphi_{X(t_1),\ldots,X(t_n)}(u_1,\ldots,u_n) = E\left(\exp\left\{i\sum_{j=1}^{n} u_j X(t_j)\right\}\right)$$

(6.4)

$$= \varphi_{X(t_1)}(u_1 + \cdots + u_n)\prod_{j=1}^{n-1}\varphi_{X(t_{j+1})-X(t_j)}(u_{j+1} + \cdots + u_n);$$

that is, the probability law is determined by the distribution of $X(t)$ and $X(t) - X(s)$, $t > s$.

Suppose $X(t)$ is an L^2 continuous additive process. Then the increments $X(t) - X(s)$ are Gaussian random variables. This can be established using characteristic functions similarly to the elementary proof for the central limit theorem. Now a *Gaussian process* is a stochastic process with all finite joint distributions Gaussian or normal. Thus if $X(t)$ is an L^2 continuous additive process on an interval J and if $X(t_0)$ is normal for some $t_0 \in J$, then $X(t)$ is a Gaussian process. For Gaussian processes, the notions of strictly stationary and wide-sense stationary are equivalent.

EXAMPLE 1.9 Let $J = [0, \infty)$. A scalar continuous L^2 process $X(t)$ with parameter space J, having stationary independent increments and satisfying

(i) $P(X(0) = 0) = 1$
(ii) $E(X(t)) = 0$, all $t \in J$
(iii) $V(X(t)) = t$, all $t \in J$

is called a *(standard) Brownian motion* or *Wiener process*. Denote this process by $W(t)$. $W(t)$ is a wide-sense stationary Gaussian process with joint characteristic function

$$\varphi_{W(t_1),\ldots,W(t_n)}(u_1,\ldots,u_n)$$

$$= \exp\left\{i\left(-\frac{1}{2}\left[t_1 v_1^2 + \sum_{j=2}^{n}(t_j - t_{j-1})v_j^2\right]\right)\right\}, \quad \text{where } v_j = \sum_{k=j}^{n} u_k.$$

The increment $W(t+1) - W(t)$ over any unit interval $[t, t+1]$ is the stochastic limit of scaled random walks $S_n/\sqrt{n}$, discussed in Example 1.8. In particular, suppose independent steps of length $\pm\sqrt{\Delta t}$ are taken at time intervals of length Δt, with

$$X_i = \begin{cases} +1 & \text{for a } +\sqrt{\Delta t} \text{ step} \\ -1 & \text{for a } -\sqrt{\Delta t} \text{ step} \end{cases}$$

where the X_i are the independent Bernoulli random variables defined in Example 1.8. Define

$$Y(t) = \sqrt{\Delta t}(X_i + \cdots + X_{[t/\Delta t]})$$

where $[\cdot]$ denotes the greatest integer $\leq \cdot$. Then

$$E(Y(t)) = 0, \qquad V(Y(t)) = \Delta t[t/\Delta t] \longrightarrow t$$

as $\Delta t \to 0$, and the central limit theorem implies that $Y(t) \to W(t)$ as $\Delta t \to 0$ in the distribution sense.

1.7 MARKOV PROCESSES

A stochastic dynamical system satisfies the Markov property (formulated by A.A. Markov in 1906) if the probable (future) state of the system at any time $t > s$ is independent of the (past) behavior of the system at times $t < s$, given the present state at time s. This is the stochastic analogue of an important property shared with solutions of initial value problems involving ordinary differential equations, and so stochastic processes satisfying this property arise naturally. In this section the Markov property is made precise and the basic properties of Markov processes are given.

Let $X(t)$ be a stochastic process with state space R^n and index set $J = [t_0, T] \subseteq [0, \infty)$. (Here the case $T = \infty$ may be considered, i.e., $J = [t_0, \infty)$.) For $t_1, t_2 \in J$ with $t_1 \leq t_2$, define

$$\mathcal{A}([t_1, t_2]) = \mathcal{A}(X(t), t_1 \leq t \leq t_2) \tag{7.1}$$

$\mathcal{A}([t_1, t_2])$ is generated by all sets of the form

$$\{\omega : X(s_1, \omega) \in B_1, \ldots, X(s_m, \omega) \in B_m\}$$
$$= \{X(s_1) \in B_1, \ldots, X(s_m) \in B_m\},$$

$t_1 \leq s_1 < s_2 < \cdots < s_m \leq t_2$, $B_i \in \mathcal{B}^n$, and m any positive integer. $X(t)$ is called a *Markov process* if the following *Markov property* holds: For $t_0 \leq s \leq t \leq T$ and all $B \in \mathcal{B}^n$

$$P(X(t) \in B | \mathcal{A}([t_0, s])) = P(X(t) \in B | X(s)) \tag{7.2}$$

with probability 1.

The following are each equivalent to the Markov property.

For $t_0 \leq s \leq t \leq T$ and $A \in \mathcal{A}([t,T])$

$\qquad P(A \mid \mathcal{A}([t_0,s])) = P(A \mid X(s))$.

For $t_0 \leq s \leq t \leq T$ and $Y, \mathcal{A}([t,T])$-integrable

$\qquad E(Y \mid \mathcal{A}([t_0,s])) = E(Y \mid X(s))$.

For $t_0 \leq t_1 \leq t \leq t_2 \leq T$, $A_1 \in \mathcal{A}([t_0,t_1])$ and $A_2 \in \mathcal{A}([t_2,T])$

$\qquad P(A_1 \cap A_2 \mid X(t)) = P(A_1 \mid X(t))P(A_2 \mid X(t))$.

For $n \geq 1$, $t_0 \leq t_1 < \cdots < t_n < t \leq T$ and $B \in \mathcal{B}_n$

$\qquad P(X(t) \in B \mid X(t_1),\ldots,X(t_n)) = P(X(t) \in B \mid X(t_n))$.

(7.3)

The conditional probability

$$P(s, X(s), t, B)) = P(X(t) \in B \mid X(s)) \qquad (7.4)$$

as a function of four variables, with $t_0 \leq s \leq t \leq T$, satisfies the following properties, where $x \in R^n$ and $B \in \mathcal{B}^n$:

For fixed s, t, and x, $P(s,x,t,\cdot)$ is a probability on $\mathcal{B}^n$.

For fixed s, t, and B, $P(s,\cdot,t,B)$ is $\mathcal{B}^n$-measurable.

For fixed s, t, u, with $s \leq u \leq t$, B and all x except possibly for a P_X-null set

$$P(s,x,t,B) = \int_{R^n} P(u,y,t,B)P(s,x,u,dy) \qquad (7.5)$$

(Chapman-Kolmogorov equation)

Actually $P(s,x,t,B)$ can be modified in such a way that the Chapman-Kolmogorov equation is satisfied for all $x \in R^n$, and such that

$$P(s,x,s,B) = I_B(x), \qquad \text{all } B \in \mathcal{B}^n.$$

In this case a function P satsifying (7.5) is called a *transition probability* or *transition function* and the notation

$$P(s,x,t,B) = P(X(t) \in B \mid X(s) = x) \qquad (7.6)$$

is employed.

The Chapman-Kolmogorov equation indicates that the transition probability can be decomposed into the state-space integral of products of probabilities to and from a location in state space, attained at an arbitrary intermediate fixed time in the parameter or index set; that is, the one-step transition probability can be written in terms of all possible combinations of two-step transition probabilities with respect to any arbitrary intermediate time.

The importance of the transition probabilities for Markov processes is that all finite-dimensional distributions of the process can be obtained from them and the initial distribution at time t_0: If $P_{X(t_0)}(A) = P(X(t_0) \in A)$, then for $t_0 \leq t_1 < \cdots < t_m \leq T, B_i \in \mathcal{B}^n$

$$P(X(t_1) \in B_1, \ldots, X(t_m) \in B_m)$$

$$= \int_{R^n} \int_{B_1} \cdots \int_{B_{m-1}} P(t_{m-1}, x_{m-1}, t_m, B_m) \tag{7.7}$$

$$\times P(t_{m-2}, x_{m-2}, t_{m-1}, dx_{m-1}) \cdots P(t_0, x_0, t_1, dx_1) P_{X(t_0)}(dx_0).$$

So this is another case where the complete probability law of the process is specified by more limited information; a probability transition function $P(s, x, t, B)$ and an initial distribution $P_{X(t_0)}$ determines uniquely (up to equivalence) a Markov process.

A Markov process with stationary increments is called *homogeneous* (with respect to time) if, for any $u > 0$,

$$P(s + u, x, t + u, B) = P(X(t + u) \in B \mid X(s + u) = x)$$

$$= P(X(t + u) - X(s + u)$$

$$\in B - x \mid X(s + u) = x) \tag{7.8}$$

$$= P(X(t) - X(s) \in B - x \mid X(s) = x)$$

$$= P(X(t) \in B \mid X(s) = x) = P(s, x, t, B).$$

Since P depends only on $t - s$, x, and B in this case, one can use the notation

$$P(s, x, t, B) = P(t - s, x, B) \tag{7.9}$$

For homogenous processes the Chapman-Kolmogorov equation (7.5) becomes

$$P(t + s, x, B) = \int_{R^n} P(s, y, B) P(t, x, dy). \tag{7.10}$$

Of course, every Markov process $X(t)$ can be transformed into a homogenous process by assuming time to be a state variable. For the transformed process $Y(t) = (t, X(t))$ the transition probability

$$Q(t, y, B) \qquad \text{for } B = C \times D, C \in \mathcal{B}([t_0, T]), D \in \mathcal{B}^n$$

is given by

$$Q(t, y, C \times D) = Q(t, (s, x), C \times D) = P(s, x, s + t, D)I_C(s + t)$$

$$(7.11)$$

and the probability $Q(t, y, \cdot)$ is then uniquely determined on $\mathcal{B}([t_0, T]) \times \mathcal{B}^n$.

A (possibly extended) random variable $\tau : \Omega \to [t_0, T]$ is called a *Markov or stopping time* with respect to the Markov process $X(t)$, if the sets $\{\tau \geq t_0\}$ and $\{\tau \leq t\}$ are in $\mathcal{A}(t)$, for each $t \in [t_0, T]$. For a Markov time τ, the σ-algebra $\mathcal{A}(\tau)$ consists of all events $A \in \mathcal{A}(T)$ such that $A \cap \{\tau \leq t\}$ is in $\mathcal{A}(t)$, each $t \in [t_0, T]$. It is clear that $\tau \wedge s = \min\{\tau, s\}$ is a Markov time if τ is and s is constant. For a continuous Markov process $X(t)$ and a closed set F

$$\tau = \inf\{t : X(t) \in F\} \tag{7.12}$$

is a Markov time called the *(first) hitting time* of the set F or the *(first) exit time* of F^c. A continuous process $X(t)$ is called a *strong Markov process* if the Markov property holds for Markov times; that is, for any Markov time τ, and any Borel set B,

$$P(X(t + \tau) \in B \mid \mathcal{A}[t_0, \tau]) = P(X(t + \tau) \in B \mid X(\tau)) \tag{7.13}$$

if $t > 0$.

EXAMPLE 1.10   The Wiener process $W(t)$ defined in Example 1.9 is a Markov process with stationary transition probability

$$P(t, x, B) = \int_B (2\pi t)^{-n/2} e^{-|y-x|^2/2t} \, dy. \tag{7.14}$$

This process is spacewise [since $W(0) = 0$] as well as timewise homogeneous. One can calculate

$$E|W(t) - W(s)|^4 = 3(t - s)^2; \tag{7.15}$$

so according to Theorem 1.6, W has a version with continuous sample paths w.p. 1. (Taking into account the discussion in Example 1.9, this illustrates that for L^2 wide-sense stationary additive processes, the notions of mean zero Gaussian increments and continuous version are equivalent.) Assume $W(t)$ has been chosen thusly. Then $W(t)$ is a strong Markov process.

1.8 DIFFUSION PROCESSES

A Markov process $X(t)$ with index set $[t_0, T]$, state space R^n, and continuous sample paths with probability 1 is called a *diffusion process* if its transition probability $P(s, x, t, B)$ is smooth in the sense that it satisfies the following three conditions for every $s \in [t_0, T)$, $x \in R^n$, and $\varepsilon > 0$:

$$\lim_{t \downarrow s} \frac{1}{t-s} \int\limits_{|y-x|>\varepsilon} P(s, x, t, dy) = 0 \tag{8.1}$$

$$\lim_{t \downarrow s} \frac{1}{t-s} \int\limits_{|y-x|\leq\varepsilon} (y-x) P(s, x, t, dy) = a(s, x) \tag{8.2}$$

$$\lim_{t \downarrow s} \frac{1}{t-s} \int\limits_{|y-x|\leq\varepsilon} (y-x)(y-x)^T P(s, x, t, dy) = B(s, x) \tag{8.3}$$

where $a(s, x)$ and $B(s, x)$ represent well-defined R^n- and $R^{n \times n}$-valued functions respectively. These functions are called the coefficients of the diffusion process. In particular, a is referred to as the *drift vector* and B the *diffusion matrix*. B is symmetric and nonnegative definite for each s and x. Condition (8.1) implies that the probability of a transition in X exceeding ε between times s and t is $o(t-s)$. Properties (8.2) and (8.3) can be written, similarly,

$$E_{s,x}(X(t) - X(s)) = E(X(t) - X(s) \mid X(s) = x)$$
$$= a(s, x)(t-s) + o(t-s) \tag{8.4}$$

$$E_{s,x}([X(t) - X(s)][X(t) - X(s)]^T) = B(s, x)(t-s) + o(t-s) \tag{8.5}$$

The drift coefficient of a diffusion process gives the time rate of change of the conditional mean of the increment of the process. Since $E_{s,x}(X(t) - X(s)) E_{s,x}(X(t) - X(s))^T = o(t-s)$ by (8.4), the diffusion matrix represents the rate of change of the conditional covariance of the increment. As examples, uniform motion with velocity ν is a diffusion process with $a = \nu$ and $B = 0$, while at the other extreme, the standard Wiener process is a diffusion with $a = 0$ and $B = I$, the identity matrix.

The transition probability $P(s, x, t, B)$ of a diffusion process is uniquely determined by the drift and diffusion coefficients of the process, under mild regularity conditions. Let the elliptic *operator* $\mathcal{L}$ be defined by

$$\mathcal{L} = \sum_{i=1}^{n} a_i(s, x) \frac{\partial}{\partial x_i} + \frac{1}{2} \sum_{i,j=1}^{n} b_{ij}(s, x) \frac{\partial^2}{\partial x_i \partial x_j} \tag{8.6}$$

where $a = (a_i)$ and $B = (b_{ij})$ are continuous vector- and matrix-valued functions, respectively. If $X(t)$, $t \in [t_0, T]$, is a diffusion process with coefficients a and B, and $g(x)$ is any bounded measurable scalar function defined in R^n, then

$$u(s, x) = E_{s,x} g(X(t)) = \int_{R^n} g(y) P(s, x, t, dy) \tag{8.7}$$

for $t \in [t_0, T]$ fixed and $s < t$ is continuous and bounded, along with its derivatives $\partial u / \partial x_i$, $\partial^2 u / \partial x_i \partial x_j$, $1 \le i, j \le n$, and so is in the domain of $\mathcal{L}$. Furthermore, $u(s, x)$ is differentiable with respect to s and satisfies *Kolmogorov's backward equation*

$$\frac{\partial u}{\partial s} + \mathcal{L}u = 0 \tag{8.8}$$

with terminal condition

$$\lim_{s \uparrow t} u(s, x) = g(x). \tag{8.9}$$

This is called the backward equation because the differentiation is with respect to the backward variables s and x. The equation specifies $P(s, x, t, \cdot)$; if the integrals u are known for g ranging over a set of functions dense in the space of bounded continuous functions on R^n, and solutions of (8.8), (8.9) are unique for given a and B, calculation of u determines $P(s, x, t, \cdot)$. In case $P(s, x, t, \cdot)$ has a density $p(s, x, t, y)$ that is continuous with respect to s and if the derivatives $\partial p / \partial x_i$ and $\partial^2 p / \partial x_i \partial x_j$ exist and are continuous with respect to s, then p satisfies (8.8) together with the terminal condition

$$\lim_{s \uparrow t} p(s, x, t, y) = \delta(x - y) \tag{8.10}$$

where δ is Dirac's delta function, that is, p is what is referred to as a *fundamental solution* of the backward equation. If, for fixed s and x such that $s \le t$, the derivatives $\partial p / \partial t$, $\partial(a_i(t, y)p) / \partial y_i$, and $\partial^2(b_{ij}(t, y)p) / \partial y_i \partial y_j$ exist and are continuous, then $p(s, x, t, y)$ is a fundamental solution of *Kolmogorov's forward* or the *Fokker-Planck equation*

$$\frac{\partial p}{\partial t} + \mathcal{L}^* p = 0 \tag{8.11}$$

where $\mathcal{L}^*$ is the formal adjoint of $\mathcal{L}$:

$$\mathcal{L}^* u = \sum_{i=1}^{n} \frac{\partial(a_i(t, y)u)}{\partial y_i} - \frac{1}{2} \sum_{i,j=1}^{n} \frac{\partial^2(b_{ij}(t, y)u)}{\partial y_i \, \partial y_j}. \tag{8.12}$$

EXAMPLE 1.11   The standard Wiener process density

$$p(s, x, t, y) = (2\pi(t - s))^{-n/2} e^{-|y - x|^2 / 2(t - s)}$$

satisfies $\partial p / \partial t = -\partial p / \partial s$. The backward equation is

$$\frac{\partial p}{\partial s} + \frac{1}{2} \sum_{i=1}^{n} \frac{\partial^2 p}{\partial x_i^2} = 0 \qquad (8.13)$$

whereas the forward equation is

$$\frac{\partial p}{\partial t} - \frac{1}{2} \sum_{i=1}^{n} \frac{\partial^2 p}{\partial y_i^2} = 0. \qquad (8.14)$$

In this case, one notices that equations (8.13) and (8.14) are identical with x and y switched.

Solutions of stochastic differential equations are Markov diffusion processes. This fact is verified in Chapter 3. The Wiener process, discussed in this chapter in Examples 1.6 to 1.8, is the prototype of such processes, and the solution of the simplest stochastic differential equation. In the next section additional properties of the Wiener process are given, including a discussion of its "derivative," Gaussian white noise.

1.9   WIENER PROCESS AND WHITE NOISE

The *standard m-dimensional Wiener process* $W(t) = \{W_1(t), \ldots, W_m(t)\}$, defined for $t \geq 0$, has R^m as its state space and is a stochastic process whose components $W_j(t)$, $j = 1, \ldots, m$, are independent scalar standard Wiener processes as defined in Examples (1.7) to (1.9) in this chapter. In particular, each W_j is a scalar process with independent, stationary, and $\mathcal{N}(0, |t - s|)$-distributed increments $W_j(t) - W_j(s)$ and satisfying $W_j(0) = 0$ w.p. 1. One can calculate

$$E|W(t) - W(s)|^4 = (m^2 + 2m)(t - s)^2, \qquad (9.1)$$

which extends (7.15) to this vector case. The process W is continuous. Theorem 1.7 indicates that this process, then, is separable and measurable. The process W is also a homogenous Markov diffusion process. A theorem of Doob, stated next, provides a criterion for determining when a continuous process is a Wiener process.

THEOREM 1.8   Suppose a process $X(t)$ is defined and continuous with probability 1 for $t \geq 0$, $X(0) = 0$ almost surely, and for $t \geq 0$ the σ-algebras $\mathcal{A}_t$ satisfy:

$$\mathcal{A}_{t_1} \subseteq \mathcal{A}_{t_2}, \qquad \text{if } t_1 < t_2.$$

Assume also that for all $t \geq s \geq 0$,

   (i) $X(t)$ is $\mathcal{A}_t$-measurable.

   (ii) $E([X(t) - X(s)] \mid \mathcal{A}_s) = 0$ w.p. 1. $\qquad\qquad\qquad\qquad$ (9.2)

   (iii) $E([X(t) - X(s)]^2 \mid \mathcal{A}_s) = t - s$ w.p. 1.

Then $X(t)$ is a Wiener process.

   In this section other sample path properties of the Wiener process are stated, and the relationship between this process and the (generalized) stochastic process known as white noise is discussed.

   For the scalar Wiener process, first of all,

$$\operatorname*{limit}_{t \to \infty} \frac{W(t)}{t} = 0, \tag{9.3}$$

analogously to the strong law of large numbers. However,

$$\limsup_{t \to \infty} \frac{W(t)}{\sqrt{2t \log\log t}} = 1 \tag{9.4}$$

and

$$\liminf_{t \to \infty} \frac{W(t)}{\sqrt{2t \log\log t}} = -1 \tag{9.5}$$

constituting the law of the iterated logarithm, gives the asymptotic behavior more precisely. For the m-dimensional Wiener process,

$$\limsup_{t \to \infty} \frac{|W(t)|}{\sqrt{2t \log\log t}} = 1 \tag{9.6}$$

   Since $tW(t)$ is a Wiener process whenever $W(t)$ is, (9.4) and (9.5) can be written as the so-called local law of the iterated logarithm:

$$\limsup_{t \downarrow 0} \frac{W(t)}{\sqrt{2t \log\log 1/t}} = 1 \tag{9.7}$$

$$\liminf_{t \downarrow 0} \frac{W(t)}{\sqrt{2t \log\log 1/t}} = -1; \tag{9.8}$$

and, for the m-dimensional process, (9.6) becomes

$$\limsup_{t \downarrow 0} \frac{|W(t)|}{\sqrt{2t \log\log 1/t}} = 1 \tag{9.9}$$

Equations (9.3) to (9.9) are all valid with probability 1.

That almost all sample paths of the Wiener process are nowhere differentiable functions was proved by Norbert Wiener. To see this consider the scalar case, and note that, for fixed nonnegative s and nonnegative t, $W(t+s) - W(s)$ is a Wiener process; then observe that (9.7) and (9.8) imply that, for $0 < \varepsilon < 1$, and for almost every sample path, there are sequences of values for h (dependent on the path) tending to zero such that

$$\frac{W(t+h) - W(t)}{h} \geq (1 - \varepsilon)\sqrt{\frac{2\log\log 1/h}{h}} \qquad (9.10)$$

and

$$\frac{W(t+h) - W(t)}{h} \leq (-1 + \varepsilon)\sqrt{\frac{2\log\log 1/h}{h}}. \qquad (9.11)$$

The right-hand sides of (9.10) and (9.11) approach $+\infty$ and $-\infty$ repectively, so $[W(t+h) - W(t)]/h$ has, with probability 1, for every fixed t, the extended real line as its cluster set. Thus the paths of $W(t)$ are nowhere differentiable, since the average rates of change experience arbitrarily large fluctuations as $h \downarrow 0$ at each time t. Hence the velocity of the Brownian particle described by the Wiener process does not exist at any time t with probability 1. In fact, it can be shown that on any finite time interval, almost all Wiener process sample paths are of unbounded variation, a situation which prevents defining integrals of the form

$$\int f(t, \omega)\, dW(t, \omega) \qquad (9.12)$$

as Stieltjes integrals along sample paths. This assertion will be returned to in the following chapters.

This introductory chapter concludes now with a brief discussion of the idealized process known as white noise.

Gaussian white noise is a model for a completely random process whose individual random variables are normally distributed. As such, it is an idealization of stochastic phenomena encountered particularly in engineering systems analysis. *Gaussian white noise* is defined in applications literature as a scalar stationary Gaussian process $\mathcal{N}(t)$ for $-\infty < t < \infty$ with $E(\mathcal{N}(t)) = 0$ and a constant spectral density function $f(\lambda)$ on the entire real line, that is, if $C(t) = E(\mathcal{N}(s)\mathcal{N}(t+s))$

$$f(\lambda) = \frac{1}{2\pi} \int_{-\infty}^{\infty} e^{-i\lambda t} C(t)\, dt = \frac{C}{2\pi}, \qquad \lambda \in R, \qquad (9.13)$$

for some constant C.

Since the spectral density function may be interpreted as measuring the relative contribution of frequency λ to the oscillatory make-up of $C(t)$, the last equation yielding that all frequencies are present equally justifies the name "white" noise in analogy with white light. Also, (9.13) implies that the covariance function $C(t) = \delta(t)$, Dirac's delta function; and so the process is uncorrelated at distinct times, and hence is independent at distinct times since it is Gaussian. In particular,

$$C(0) = \int_{-\infty}^{\infty} f(\lambda) \, d\lambda = \infty.$$

The nature of the covariance $C(t)$ indicates that such a process cannot be realized, and that white noise is not a stochastic process in the usual sense. A rigorous treatment of such processes can be carried out in the setting of generalized functions or Schwartz distributions. The relationship between Gaussian white noise $\mathcal{N}(t)$ and (scalar) standard Wiener process $W(t)$ can be understood formally by considering the covariance of the latter

$$C_W(t,s) = E(W(t)W(s)) = \min(t,s). \tag{9.14}$$

When it exists the covariance of the derivative process satisfies

$$C_{\dot{W}}(t,s) = \frac{\partial^2}{\partial s \partial t} C_W(t,s),$$

so in this case

$$C_{\dot{W}}(t,s) = \frac{\partial^2}{\partial s \partial t} \min(t,s) = \delta(t-s) = C(t-s); \tag{9.15}$$

that is, the covariance of the derivative of the Wiener process is the covariance of white noise.

It is emphasized that the calculation in (9.15) is formal; it has already been established that the Wiener process is nowhere differentiable. Nevertheless, the preceding argument can be made rigorous within the framework of distribution theory. It is in this generalized sense that Gaussian white noise is the derivative of the Wiener process. Multivariate white noise has the same relationship with the multidimensional Wiener process.

2

Stochastic Integrals and Ito's Formula

2.0 INTRODUCTION

In recent years, it has become apparent that physical systems, classically modeled by deterministic differential equations, can be more satisfactorily modeled by certain stochastic counterparts if random effects in the physical phenomena as well as measuring devices are to be taken into account. In this context, an ordinary differential equation

$$\frac{dx}{dt} = f(t, x) \tag{0.1}$$

would be replaced by a random differential equation

$$\frac{dX}{dt} = F(t, X, Y), \tag{0.2}$$

where $Y = Y(t)$ represents some stochastic input process explicitly. Of course, (0.2) is a more complicated equation from two points of view. A solution is a stochastic process, that is, an indexed (over the probability space) family of functions of the time variable rather than a single such function. Secondly, if the sample path structure of the input process Y is sufficiently pathological (e.g., if the sample paths are not integrable), then equation (0.2) must be reinterpreted. In particular, it may not be possible to interpret (0.2) as an ordinary differential equation along each path. This happens when (0.2) has the form

$$\frac{dX}{dt} = f(t, X) + g(t, X) \mathcal{N} \tag{0.3}$$

with $\mathcal{N}$ being a Gaussian white noise process. Equation (0.3) has been popular in the engineering literature especially, since Gaussian white noise approximates the effect of the superposition of a large number of small random disturbances, a situation encountered in engineering systems. However, the irregularity of the sample paths of $\mathcal{N}$ makes (0.3) intractable mathematically. Just as a solution $x(t)$ of (0.1) satisfies the deterministic integral equation

$$x(t) = x(t_0) + \int_{t_0}^{t} f(s, x(s)) \, ds$$

a solution of (0.3) should be a solution of the random integral equation

$$X(t) = X(t_0) + \int_{t_0}^{t} f(s, X(s)) \, ds + \int_{t_0}^{t} g(s, X(s)) \mathcal{N}(s) \, ds; \tag{0.4}$$

but unfortunately the last integral in (0.4) cannot be defined in any meaningful way. To deal with this difficulty, the integral in question is replaced by an integral of the form

$$\int_{t_0}^{t} g(s, X(s)) \, dW(s) \tag{0.5}$$

where W is the Wiener process, and with the motivation being that, at least formally, $\dot{W}(t) = \mathcal{N}(t)$ and so $dW(t) = \mathcal{N}(t) \, dt$. However, even equation (0.4) with the second integral replaced by (0.5) is not unequivocal. The natural interpretation of (0.5) would be as a Stieltjes integral along sample paths. This is not possible, since the sample paths of the Wiener process are not of bounded variation as shown in the next section. [See Rudin (1976), for example, for a discussion of ordinary Riemann-Stieltjes integrals.] Furthermore, if different choices of the τ_i are made from the subinterval $[t_{i-1}, t_i]$, the natural approximating sums

$$\sum_{i=1}^{n} g(\tau_i, X(\tau_i))[W(t_i) - W(t_{i-1})] \tag{0.6}$$

converge in the mean square sense to different values of the integral. If, in (0.6), τ_i is taken as t_{i-1}, the Ito integral results. The impact is that the integral inherits many of the probabilistic properties of the Wiener process at the expense of the corresponding calculus differing from what one would expect in the Stieltjes case. The main instrument of this calculus is Ito's formula, which can be thought of as a probabilistic chain rule; it yields a stochastic equation for any sufficiently smooth function of a solution. As such, Ito's formula is the basis for any analysis of solutions of stochastic equations along the lines of the qualitative approaches used in ordinary differential and integral equations.

The object in this chapter is to prove the simplest form of Ito's theorem, which is the statement of this formula, after introducing the Ito integral for a class of random functions $g(t)$. Also verified, for a fairly general such class of random functions, is that the stochastic integral

$$\int_a^b g(t)\, dW(t)$$

of Ito satisfies

$$E \int_a^b g(t)\, dW(t) = 0, \tag{0.7}$$

$$E \left| \int_a^b g(t)\, dW(t) \right|^2 = \int_a^b E|g(t)|^2\, dt, \tag{0.8}$$

and, viewed as a function of the upper limit of integration, forms a martingale. The martingale property will be discussed in Section 2.3. These properties are natural extensions of the facts that: $W(t)$ is a martingale, $E[W(b) - W(a)] = 0$, and $E|W(b) - W(a)|^2 = b - a$, and illlustrate the tractability that motivates the choice of the Ito integral from a mathematical point of view. (In Chapter 6, it will be shown that the Ito integral provides an approximation to relevant but more mathematically pathological random process models of physical phenomena—which suggests the choice of Ito calculus from an applications standpoint.) The next section opens the discussion by considering the problem of defining integrals of the form (0.5).

2.1 NONANTICIPATING FUNCTIONS

Integrals of the form

$$\int_a^b g(t)\, dW(t) \tag{1.1}$$

can be defined, in the natural way, starting with the approximations

$$\sum_{i=1}^n g(\tau_i)[W(t_i) - W(t_{i-1})] \tag{1.2}$$

with τ_i chosen arbitrarily in the subinterval $[t_{i-1}, t_i]$ of the partition $0 \le a = t_0 < t_1 < \cdots < t_n = b$, if g is a deterministic function which is Lebesque or Riemann-Stieltjes integrable. If g is a random function, analogously, one expects to start from (1.2) where, for each i, $g(\tau_i)$ denotes a random variable. The difficulty alluded to in the previous section arises from the fact that the random variable $g(\tau_i)$ may not be measurable with respect to the σ-algebra generated by $W(t_{i-1})$. This situation is illustrated in this section by attempting to define the integral

$$\int_a^b W(t)\, dW(t) \tag{1.3}$$

where $W(t)$, $t \in [0, \infty)$ is a standard 1-dimensional Wiener process.

To begin, the next result has as a consequence that $W(t)$ is of unbounded variation on any finite time interval with probability 1; therefore, defining (1.3) as a Stieltjes integral is not possible. In the following, assume $0 \le a = t_0^{(n)} < \cdots < t_n^{(n)} = b$ denotes a sequence of partitions of the interval $[a, b]$, $\Delta t_k^{(n)} = t_k^{(n)} - t_{k-1}^{(n)}$, $\delta_n = \max_k \Delta t_k^{(n)}$, and $\Delta W_k^{(n)} = W(t_k^{(n)}) - W(t_{k-1}^{(n)})$.

LEMMA 2.1

$$\lim_{\delta_n \to 0} \sum_{k=1}^n (\Delta W_k^{(n)})^2 = b - a \tag{1.4}$$

where the mode of convergence is mean square. Further if $\sum \delta_n < \infty$, (1.4) holds with probability 1.

Proof: If $S_n = \sum_{k=1}^n (\Delta W_k^{(n)})^2$, then (1.4) follows immediately for convergence in the mean since $E(S_n) = b - a$, all n.

Now, since $(\Delta W_j^{(n)})^2$ and $(\Delta W_k^{(n)})^2$ are independent if $j \ne k$, one has

$$V(S_n) = \sum_{k=1}^n V(\Delta W_k^{(n)})^2$$

$$= \sum_{k=1}^n [E(\Delta W_k^{(n)})^4 - (\Delta t_k^{(n)})^2]$$

$$= 2 \sum_{k=1}^n (\Delta t_k^{(n)})^2 \le 2(b-a)\delta_n \longrightarrow 0 \qquad \text{as } \delta_n \longrightarrow 0.$$

Here (7.15) of Chapter 1 was used: $E(\Delta W_k^{(n)})^4 = 3(\Delta t_k^{(n)})^2$. Thus the first part of the lemma is verified.

Furthermore, by Fatou's lemma, we have

$$E\left(\sum_{n=1}^{\infty}(S_n - E(S_n))^2\right) = E\left(\liminf_{N\to\infty}\sum_{n=1}^{N}(S_n - E(S_n))^2\right)$$

$$\leq \liminf_{N\to\infty}\sum_{n=1}^{N} E(S_n - E(S_n))^2 = \sum_{n=1}^{\infty} V(S_n) \leq 2(b-a)\sum_{n=1}^{\infty} \delta_n. \qquad (1.5)$$

Thus, if $\sum \delta_n < \infty$, (1.5) implies that

$$P\left(\sum_{n=1}^{\infty}(S_n - E(S_n))^2 < \infty\right) = 1.$$

And so, with probability 1, $S_n \to E(S_n)$, as $n \to \infty$. This establishes the lemma. $\square$

Note that if $W(t)$ is a m-dimensional Wiener process, (1.4) has the form

$$\lim_{\delta_n \to 0} \sum_{k=1}^{n} \Delta W_k^{(n)} \Delta W_k^{(n)T} = (b-a)I \qquad (1.6)$$

where I is the $m \times m$ identity matrix. Taking the trace of both sides in (1.6) yields

$$\lim_{\delta_n \to 0} \sum_{k=1}^{n} |\Delta W_k^{(n)}|^2 = m(b-a). \qquad (1.7)$$

Now suppose, once again, δ_n is chosen so that $\sum \delta_n < \infty$, and the partition is uniform. Then

$$\sum_{k=1}^{n} |\Delta W_k^{(n)}|^2 \leq \max_{k} |\Delta W_k^{(n)}| \sum_{k=1}^{n} |\Delta W_k^{(n)}|. \qquad (1.8)$$

As $n \to \infty$, the left-hand side of (1.8) approaches $m(b-a)$, and $\max_k |\Delta W_k^{(n)}| \to 0$ due to continuity of sample paths and uniformity of the partitions. The conclusion drawn from (1.8) in this case is

$$\sum_{k=1}^{n} |\Delta W_k^{(n)}| \longrightarrow \infty \quad \text{as } n \longrightarrow \infty, \qquad (1.9)$$

which establishes the following result.

THEOREM 2.2   The sample paths of $W(t)$ are of unbounded variation on every finite time interval with probability 1.

It has been pointed out that Theorem 2.2 indicates difficulty in defining (1.3) as a Stieltjes integral. Furthermore, one would expect that sums of the form

$$\sum W(\tau_k^{(n)}) \Delta W_k^{(n)}$$

would converge to such an integral in an appropriate sense independent of the choice of $\tau_k^{(n)}$ if the integral were of Stieltjes type. However, the next result shows that the mean square limit of

$$\sum [\lambda W(t_k^{(n)}) + (1-\lambda)W(t_{k-1}^{(n)})] \Delta W_k^{(n)} \tag{1.10}$$

depends on λ.

THEOREM 2.3   As $\delta_n \to 0$, the expression (1.10) tends to $\frac{1}{2}(W^2(b) - W^2(a)) + (\lambda - \frac{1}{2})(b-a)$ in the mean square sense.

Proof: For the $\lambda = 0$ case (1.10) can be written

$$\sum_{k=1}^{n} W(t_{k-1}^{(n)}) \Delta W_k^{(n)} = \frac{1}{2}\left[W^2(b) - W^2(a) - \sum_{k=1}^{n}(\Delta W_k^{(n)})^2\right] \tag{1.11}$$

and the result follows from Lemma 2.1. That is,

$$\sum_{k=1}^{n} W(t_{k-1}^{(n)}) \Delta W_k^{(n)} \longrightarrow \frac{1}{2}\left[W^2(b) - W^2(a) - (b-a)\right] \tag{1.12}$$

in the mean square sense, as $\delta_n \to 0$.

For the $\lambda = 1$ case, rearranging $W(t_k^{(n)})\Delta W_k^{(n)}$ as

$$(\Delta W_k^{(n)})^2 + W(t_{k-1}^{(n)}) \Delta W_k^{(n)}$$

and using Lemma 2.1 once again as well as (1.12), it follows that

$$\sum_{k=1}^{n} W(t_k^{(n)}) \Delta W_k^{(n)} \longrightarrow \frac{1}{2}[W^2(b) - W^2(a) + (b-a)] \tag{1.13}$$

in the mean square sense, as $\delta_n \to 0$. The theorem now follows by multiplying (1.12) by $1-\lambda$ and (1.13) by λ, so that (1.10) is obtained. $\square$

So Theorems 2.2 and 2.3 demonstrate some problems in defining a stochastic integral

$$\int_a^b Y(t)\,dW(t),$$

for a class of processes $\{Y(t)\}$ large enough to include the Wiener process, by simply constructing an ordinary Stieltjes type integral for each sample path, that is, by treating the probability space variable as a parameter. This situation requires a new type of integral whose calculus one might expect would differ from the Riemann-Stieltjes calculus. In this book, one such integral is discussed from the point of view of differential equations. It should be mentioned, at this point, that developing an integral general enough to include both ordinary Stieltjes integrals as well as stochastic integrals of the type to be discussed here constitutes a very active research area currently. [McShane (1974) constitutes one such recent attempt.]

How then should such an integral be defined? As indicated at the outset of this section, it seems reasonable to start by defining the integral for random step functions. However, Theorem 2.3 demonstrates that there are infinitely many (indexed by the parameter λ) such definitions for random step functions leading to distinct mean square stochastic integrals. Specifically, for any fixed λ, $0 \leq \lambda \leq 1$, the random step function approximation of $W(t)$

$$\varphi_n^\lambda(t) = \lambda W(t_k^{(n)}) + (1 - \lambda)W(t_{k-1}^{(n)}), \qquad t_{k-1}^{(n)} \leq t < t_k^{(n)}$$

for which

$$\int_a^b \varphi_n^\lambda(t)\,dW(t) = \sum_{k=1}^n \varphi_n^\lambda(t_{k-1})\,\Delta W_k^{(n)} \tag{1.14}$$

results in the stochastic integral (by taking mean square limit as $\delta_n \to 0$)

$$\int_a^b W(t)\,dW(t) = \tfrac{1}{2}[W^2(b) - W^2(a)] + \left(\lambda - \tfrac{1}{2}\right)(b - a). \tag{1.15}$$

For example, when $\lambda = \tfrac{1}{2}$ the Stratonovich stochastic integral results. (This integral will be discussed in more detail in Chapter 3.) In this case the form of (1.15) indicates that the corresponding stochastic calculus will be analogous to the ususal Riemann-Stieljes calculus. However, a disadvantage of this and any $\lambda > 0$ case is that the expression

$$\lambda W(t_k^{(n)}) + (1 - \lambda)W(t_{k-1}^{(n)}) \tag{1.16}$$

is not measurable with respect to the σ-algebra generated by $W(t_{k-1}^{(n)})$, a situation which prevents the martingale property as well as (0.7) and (0.8) from holding. The corresponding step functions φ_n^λ are said to be *anticipating* in this case. The *nonanticipating* case, namely, when (1.16) is measurable with respect to the σ-algebra generated by $W(t_{k-1}^{(n)})$ and hence independent of the increment $\Delta W_k^{(n)}$, occurs only when $\lambda = 0$. This leads to the Ito case, and with the random step functions

$$\varphi_n^0(t) = W(t_{k-1}^{(n)}), \qquad t_{k-1}^{(n)} \le t < t_k^{(n)}$$

in (1.14), one obtains the special case of (1.15),

$$\int_a^b W(t)\, dW(t) = \tfrac{1}{2} \left\{ [W^2(b) - W^2(a)] - (b-a) \right\}.$$

To conclude this section properties (0.7) amd (0.8) are verified for the Ito definition of the stochastic integral (1.3).

THEOREM 2.4   The integral (1.3) interpreted in the Ito sense satisfies

(a) $E \displaystyle\int_a^b W(t)\, dW(t) = 0$ (1.17)

and

(b) $E \left| \displaystyle\int_a^b W(t)\, dW(t) \right|^2 = \tfrac{1}{2}(b^2 - a^2).$ (1.18)

Proof: Since $W(t_{k-1}^{(n)})$ and $\Delta W_k^{(n)}$ are independent with mean zero for each k and n, one has for each of the approximating sums for (1.3)

$$E \sum W(t_{k-1}^{(n)})\, \Delta W_k^{(n)} = \sum EW(t_{k-1}^{(n)}) E\, \Delta W_k^{(n)} = 0$$

and so (1.17) follows proceeding to the limit in the mean square sense. Writing

$$\left[\sum W(t_{k-1}^{(n)})\, \Delta W_k^{(n)} \right]^2 = \sum [W(t_{k-1}^{(n)})]^2 [\Delta W_k^{(n)}]^2$$
$$+ 2 \sum_{k<j} W(t_{k-1}^{(n)})\, \Delta W_k^{(n)} W(t_{j-1}^{(n)})\, \Delta W_j^{(n)}$$

and noting that for $k < j$, $W(t_{k-1}^{(n)})\, \Delta W_k^{(n)} W(t_{j-1}^{(n)})$ and $\Delta W_j^{(n)}$ are independent with the latter having mean zero, one has

$$E \left[\sum W(t_{k-1}^{(n)})\, \Delta W_k^{(n)} \right]^2 = \sum E[W(t_{k-1}^{(n)})]^2 E[\Delta W_k^{(n)}]^2$$
$$= \sum t_{k-1}^{(n)} \Delta t_k^{(n)}.$$

Since this expression converges to $\int_a^b t\, dt$ as $\delta_n \to 0$, (1.18) is verified. $\square$

Finally, note that, for any λ,

$$E \sum_{k=1}^{n} [\lambda W(t_k^{(n)}) + (1 - \lambda)W(t_{k-1}^{(n)})] \Delta W_k^{(n)} = \lambda(b - a). \qquad (1.19)$$

Therefore (1.17) holds precisely when $\lambda = 0$, that is, only in the Ito case.

The next section outlines the definition of the Ito integral for a general class of nonanticipating random functions.

2.2   DEFINITION OF THE ITO STOCHASTIC INTEGRAL

This section illustrates a basic attribute of the Ito integral which mathematically motivates its choice, namely, that it can be defined for a fairly general class of nonanticipating random functions in such a way as to preserve basic Wiener process properties. The properties for developing the integral follows the lines of general Lebesgue-type integration theory. [See Bartle (1966), for example—specifically for the integration theory results required in Lemma 2.5.] In particular, for a fixed parameter interval, the integral is defined as a continuous linear random functional, that is, a linear mapping of a space of random functions into a space of random variables. The spaces involved are at least complete metric spaces, and a class of elementary functions in the domain space, in this case, the random step functions, is designated. This integral is first defined for this class and then using continuity and linearity extended to the entire proposed domain. To begin, the spaces involved are described.

Let $W(t)$ be a standard Wiener process defined on the probability space $(\Omega, \mathcal{A}, P)$ and let $\{\mathcal{A}(t) : t \in [a,b]\}$ be a family of sub-σ-algebras of $\mathcal{A}$ satisfying the following conditions

(i) $\mathcal{A}(t_1) \subseteq \mathcal{A}(t_2)$ if $t_1 < t_2$.
(ii) $W(t)$ is $\mathcal{A}(t)$-measurable.
(iii) For $s > 0$, $W(t + s) - W(t)$ is independent of $\mathcal{A}(t)$.

Note that the σ-algebras $\mathcal{W}[a, t]$ generated by the Wiener process itself up to time t satisfy the requirements for the family $\mathcal{A}(t)$; in some cases, however, it is convenient to work with $\mathcal{A}(t)$ larger than $\mathcal{W}[a, t]$. Note also that if, for a random function $f(t, \omega)$ the random variables $f(t, \cdot)$ are $\mathcal{A}(t)$-measurable, for almost all t, then f is nonanticipating.

Let $\mathcal{L}^2$ denote the class of random functions $f(t, \omega)$ satisfying the conditions

(a) f is measurable on $[a, b] \times \Omega$.

(b) $f(t, \cdot)$ is $\mathcal{A}(t)$-measurable, for almost all $t \in [a, b]$.
(c) $\int_a^b E f^2(t, \omega) \, dt < \infty$.

Further let $\mathcal{E}$ denote the subclass of $\mathcal{L}^2$ consisting of the random step functions, that is, $\mathcal{E}$ is the class of random functions $f(t, \omega)$ satisfying (a), (b), (c) above together with

(d)   There is a partition $a = t_0 < t_1 < \cdots < t_n = b$ such that $f(t, \cdot) = f(t_i, \cdot), t_i \leq t < t_{i+1}$.

The space $\mathcal{L}^2$ equipped with the norm

$$\|f\| = \left[\int_a^b E f^2(t, \omega) \, dt \right]^{1/2} \tag{2.1}$$

forms a real Hilbert space, and

LEMMA 2.5 $\mathcal{E}$ is dense in $\mathcal{L}^2$.

Proof: Proving the lemma requires, for a given $f \in \mathcal{L}^2$, exhibiting a sequence $\{f_n\} \subseteq \mathcal{E}$ such that

$$\|f_n - f\| \longrightarrow 0 \text{ as } n \longrightarrow \infty. \tag{2.2}$$

First of all, by setting $f(t) = 0$, $t \notin [a, b]$ f can be considered defined on the entire line $(-\infty, \infty)$. Then one has, from Fubini's theorem and (c),

$$E \int_{-\infty}^{\infty} f^2(t) \, dt = \int_{-\infty}^{\infty} E f^2(t) \, dt < \infty, \tag{2.3a}$$

so that, w.p. 1,

$$\int_{-\infty}^{\infty} f^2(t) \, dt < \infty. \tag{2.3b}$$

Now it follows by two applications of the dominated convergence theorem that

$$E \int_{-\infty}^{\infty} |f(t + h) - f(t)|^2 \, dt \longrightarrow 0 \qquad \text{as } h \longrightarrow 0. \tag{2.4}$$

To see this, first note that

$$\int_{-\infty}^{\infty} |f(t + h) - f(t)|^2 \, dt \leq 4 \int_{-\infty}^{\infty} f^2(t) \, dt; \tag{2.5}$$

so (2.3b) and (2.5) imply

$$\int_{-\infty}^{\infty} |f(t + h) - f(t)|^2 \, dt \longrightarrow 0 \qquad \text{as } h \longrightarrow 0 \quad \text{w.p. 1}, \tag{2.6}$$

by the dominated convergence theorem for Lebesgue integration along sample paths. Now, from (2.5),

$$E \int_{-\infty}^{\infty} |f(t+h) - f(t)|^2 \, dt \leq 4E \int_{-\infty}^{\infty} f^2(t) \, dt,$$

and so (2.3a) permits the dominated convergence theorem to be applied in the probability space yielding (2.4).

Define the functions $\varphi_j(t) = [jt]/j$, j an integer ≥ 1 where $[\ ]$ denotes the "greatest integer in" function. Since $\varphi_j(t) \to t$ as $j \to \infty$, (2.6) can be replaced by

$$E \int_{-\infty}^{\infty} |f(s + \varphi_j(t)) - f(s+t)|^2 \, ds \longrightarrow 0 \qquad \text{as } j \longrightarrow \infty \qquad (2.7)$$

for any fixed t. As this expression is bounded by $4E \int_{-\infty}^{\infty} f^2(s) \, ds < \infty$, the dominated convergence theorem can be applied, which together with Fubini's theorem, yields

$$\int_{-\infty}^{\infty} \int_{a-1}^{b} E|f(s + \varphi_j(t)) - f(s+t)|^2 \, dt \, ds$$

$$= \int_{a-1}^{b} E \int_{-\infty}^{\infty} |f(s + \varphi_j(t)) - f(s+t))|^2 \, ds \, dt \longrightarrow 0$$

$$\text{as } j \longrightarrow \infty. \quad (2.8)$$

Therefore there exists a subsequence $\{j_n\}$ such that

$$\int_{a-1}^{b} E|f(s + \varphi_{j_n}(t)) - f(s+t)|^2 \, dt \longrightarrow 0$$

$$\text{as } j_n \longrightarrow \infty, \text{ for almost all } s. \quad (2.9)$$

Fixing such an $s \in [0,1]$ and replacing t by $t - s$ in (2.9), one has

$$\int_{a-1+s}^{b+s} E|f(s + \varphi_{j_n}(t-s)) - f(t)|^2 \, dt \longrightarrow 0. \qquad (2.10)$$

Therefore, since $0 \leq s \leq 1$, (2.10) implies

$$\int_{a}^{b} E|f(s + \varphi_{j_n}(t-s)) - f(t)|^2 \, dt \longrightarrow 0. \qquad (2.11)$$

Taking $f_n(t) = f(s + \varphi_{j_n}(t-s))$, one has that $f_n \in \mathcal{E}$ and (2.11) states that $\|f_n - f\| \to 0$ as $n \to \infty$, and the proof is complete. $\square$

Denote by $L^2(\Omega)$ the usual Hilbert space of random variables F on $(\Omega, \mathcal{A}, P)$ with finite second moments,

$$\|F\|_2 = \{E|F|^2\}^{1/2} < \infty.$$

The Ito integral

$$I(f) = \int_a^b f(t)\, dW(t) \tag{2.12}$$

can now be defined as a linear mapping,

$$I : \mathcal{L} \longrightarrow L^2$$

starting from the definition of I on $\mathcal{E}$ given by

$$I(f) = \sum_{k=1}^n f(t_{k-1})\, \Delta W_k, \tag{2.13}$$

where $f(t) = f(t_{k-1})$, $t_{k-1} \le t < t_k$, is in $\mathcal{E}$ and $\Delta W_k = W(t_k) - W(t_{k-1})$. The next theorem summarizes the situation and constitutes the main result of this section.

THEOREM 2.6   The integral I defined on $\mathcal{E}$ by (2.13) extends as a continuous linear random functional from $\mathcal{L}^2$ into L^2 which satisfies

(a) $E(I(f)) = 0$.
(b) $\|I(f)\|_2 = \|f\|$.

Proof: Let $f(t) = f(t_{k-1})$, $t_{k-1} \le t < t_k$, be in $\mathcal{E}$. Linearity of I is clear.

$$\|I(f)\|_2^2 = E([I(f)]^2) = E\left\{ \left[\sum_{k=1}^n f(t_{k-1})\, \Delta W_k \right]^2 \right\}. \tag{2.14}$$

Using the independence of the increments ΔW_k as well as the independence of $f(t_{k-1})$ and ΔW_k, and the fact that $E\,\Delta W_k = 0$, (2.14) can be written

$$\|I(f)\|_2^2 = \sum_{k=1}^n E(f^2(t_{k-1})) E([\Delta W_k]^2)$$

$$= \sum_{k=1}^n E(f^2(t_{k-1}))\, \Delta t_k = \|f\|^2. \tag{2.15}$$

Thus I is a continuous linear mapping from $\mathcal{E}$ into L^2. Also, that $E(I(f)) = 0$ is clear from the independence and mean increment zero conditions mentioned above.

Let $f \in \mathcal{L}^2$, and choose $\{f_n\} \subseteq \mathcal{E}$ such that $f_n \to f$. The sequence $\{f_n\}$ is a Cauchy sequence in $\mathcal{L}^2$; so, by linearity and continuity of $\mathcal{I}$, the sequence $\{\mathcal{I}(f_n)\}$ is a Cauchy sequence in the complete space L^2. Defining

$$\mathcal{I}(f) = \underset{n \to \infty}{\text{limit}}\ \mathcal{I}(f_n)$$

extends the definition of $\mathcal{I}$ to the entire space $\mathcal{L}^2$. Properties (a) and (b) are preserved in the limiting process, and therefore are inherited from the analogous properties for step functions, and this completes the proof. $\square$

Note that (a) and (b) are precisely the extensions of properties (1.17) and (1.18), respectively, which were verified for the Ito integral

$$\int_a^b W(t)\, dW(t)$$

in Theorem 2.4 of the last section.

Actually the Ito integral can be defined for a larger class of nonanticipating random functions than $\mathcal{L}^2$ at the expense of losing properties (a) and (b). To conclude this section, an outline of the procedure is now given; details may be found in Ito (1961).

Let $\mathcal{L}_s$ represent the class of random functions f which satisfy the defining conditions (a) and (b) for the class $\mathcal{L}^2$ given at the beginning of this section and

$$(c_s) \qquad \int_a^b f^2(t)\, dt < \infty \quad \text{w.p. 1.}$$

Since the corresponding property for $\mathcal{L}^2$

$$(c) \qquad \int_a^b E f^2(t)\, dt < \infty$$

implies (c_s), $\mathcal{L}^2 \subseteq \mathcal{L}_s$. In $\mathcal{L}_s$ convergence in the sense of the metric

$$\|f - g\|_s = E \left(\frac{\left[\int_a^b |f(t) - g(t)|^2\, dt\right]^{1/2}}{1 + \left[\int_a^b |f(t) - g(t)|^2\, dt\right]^{1/2}} \right)$$

is equivalent to convergence in probability for

$$\int_a^b |f(t) - g(t)|^2\, dt;$$

$\mathcal{E}$ is dense in $\mathcal{L}_s$ in this sense. Also, for $f \in \mathcal{L}^2$,

$$\|f\|_s \leq \|f\|_2.$$

Now denote by S the collection of all random variables X on the probability space $(\Omega, \mathcal{A}, P)$. As mentioned in Section 1.4. convergence in S in the sense of the metric

$$\|X - Y\|_S = E\left(\frac{|X - Y|}{1 + |X - Y|}\right)$$

is equivalent to convergence in probability or stochastic convergence in S. If $X \in L^2(\Omega, \mathcal{A}, P)$,

$$\|X\|_S \leq \|X\|_2.$$

The Ito integral, defined by (2.13) as a mapping

$$I : \mathcal{E} \longrightarrow L^2,$$

and extended to a mapping from $\mathcal{L}^2$ into L^2 in this section, can be further extended, by means of continuity and linearity, to a mapping between the metric spaces $\mathcal{L}_s$ and S.

2.3 STOCHASTIC INTEGRALS AS FUNCTIONS OF THE UPPER LIMIT

If $f \in \mathcal{L}_s$, and I_B is the indicator function corresponding to the arbitrary Borel set $B \subseteq [a, b]$, then

$$fI_B \in \mathcal{L}_s. \tag{3.1}$$

Therefore, the stochastic integral $\int_B f(t)\, dW(t)$ is defined by

$$\int_B f(t)\, dW(t) = I(fI_B) = \int_a^b fI_B(t)\, dW(t) \tag{3.2}$$

and satisfies the additive property

$$\int_{B_1 \cup B_2} f(t)\, dW(t) = \int_{B_1} f(t)\, dW(t) + \int_{B_2} f(t)\, dW(t) \tag{3.3}$$

for any disjoint Borel subsets B_1 and B_2 of $[a, b]$. In particular, if $a \leq t_1 \leq t_2 \leq b$

$$\int_a^{t_2} f(t)\, dW(t) = \int_a^{t_1} f(t)\, dW(t) + \int_{t_1}^{t_2} f(t)\, dW(t). \tag{3.4}$$

Setting $X(t) = \int_a^t f(s)\, dW(s)$, (3.4) becomes

$$X(t_2) = X(t_1) + \int_{t_1}^{t_2} f(s)\, dW(s);$$

it is clear that $X(t)$ is a Markov process. Furthermore if τ is a Markov time, with $\tau \leq b$ w.p. 1, $X(\tau)$ is defined by

$$\int_a^\tau f(s)\, dW(s) = \int_a^b f(s) I_{\{s < \tau\}}\, dW(s)$$

since $f(\cdot) \in \mathcal{L}_s$ implies that $f(\cdot) I_{\{\cdot < \tau\}} \in \mathcal{L}_s$. That X is a strong Markov process follows readily. Other properties of the process X are presented in this section.

The concept of *martingale*, attributed to P. Levy, is central in this discussion; a stochastic process $\{X(t) : a \leq t \leq b\}$ on a probability space $(\Omega, \mathcal{A}, P)$ is called a *martingale* with respect to the nondecreasing family $\{\mathcal{A}(t) : a \leq t \leq b\}$ of sub-σ-algebras of $\mathcal{A}$ if, for each $t \in [a, b]$, $X(t)$ is $\mathcal{A}(t)$-measurable, $E(|X(t)|) < \infty$, and for each $s > 0$,

$$E\left(X(t+s) \mid \mathcal{A}(t)\right) = X(t). \tag{3.5}$$

Such processes are sometimes referred to as "fair game" processes; if $X(t)$ represents a gambler's fortune at time t, the game is fair if his expected fortune at a future time $t + s$ given the game history up to some previous time t is precisely the fortune at time t. *Sub- (super-) martingales* are stochastic processes which satisfy the conditions in the definition above except that equality in (3.5) is relaxed to $\geq$ ($\leq$). For example, if $X(t)$ is a supermartingale, $-X(t)$ is a submartingale. Furthermore, if $X(t)$ is a martingale, and g is a convex function, then Jensen's inequality [(4.4) of Chapter 1] implies that $g(X(t))$ is a submartingale; in particular, $|X(t)|^r$, $r \geq 1$, is a submartingale. Submartingales $Y(t)$ defined on a closed interval $[a, b]$ satisfy the following key inequalities

$$P\left(\sup_{[a,b]} Y(t) > r\right) \leq E(|Y(b)|^p)/r^p \tag{3.6}$$

for arbitrary real numbers $r > 0$ and $p \geq 1$, and

$$E\left(\sup_{[a,b]} |Y(t)|^r\right) \leq \left(\frac{r}{r-1}\right)^r E(|Y(b)|^r) \tag{3.7}$$

for $r > 1$. For positive supermartingales $Y(t)$, one has

$$P\left(\sup_{[a,b]} Y(t) > r\right) \leq EY^p(a)/r^p \tag{3.6a}$$

for $r > 0$ and $p \geq 1$. Finally sub- (super-) martingales exhibit with probability 1 convergence as $t \downarrow a$ ($t \uparrow b$). These results hold on appropriate semi-infinite intervals, as well. For example, if $Y(t)$ is a positive supermartingale on $[a, \infty)$, there exists a (non-negative) random variable Y_∞

(with $EY_\infty < \infty$) such that

$$\underset{t \to \infty}{\text{limit}}\, Y(t) = Y_\infty \quad \text{w.p. 1.} \tag{3.8}$$

[See Doob, (1955), for example.]

It is clear that the Wiener process is a martingale, and the next result indicates that the martingale property extends to stochastic integrals. Kussmaul (1977) gives a detailed account of the relationship between stochastic integration and martingale theory. Attempts to generalize the stochastic integral over the past twenty years have involved defining integrals with martingales (Doob, 1955) and various generalizations of martingales replacing the Wiener process as integrators.

THEOREM 2.7   If $f \in \mathcal{L}^2$, the stochastic integral

$$X(t) = \int_a^t f(s)\, dW(s)$$

is a martingale with respect to the family of sub-σ-algebras $\{\mathcal{A}(t)\}$. (Recall that, in the definition of the space $\mathcal{L}^2$ of random functions, $\{\mathcal{A}(t)\}$ is assumed to be a nondecreasing family of sub-σ-algebras of $\mathcal{A}$ in the underlying probability space $(\Omega, \mathcal{A}, P)$ such that the Wiener process $\{W(t)\}$ is $\mathcal{A}(t)$-measurable and the increments $W(t+s) - W(t)$, $s > 0$, are independent of $\mathcal{A}(t)$.)

Proof: First, note that $E|X(t)| < \infty$ follows from $X(t) \in L^2$. For step functions f independence of the increment $X(t+s) - X(t)$ and the σ-algebra $\mathcal{A}(t)$ implies

$$E(X(t+s) - X(t)|\mathcal{A}(t)) = E(X(t+s) - X(t)) = 0$$

For general $f \in \mathcal{L}^2$, the stochastic integral $X \in L^2$, and there is a sequence of step functions $\{f_n\}$ with corresponding integrals X_n such that $f_n \to f$ in $\mathcal{L}^2$ and $X_n \to X$ in L^2. The result follows for general $f \in \mathcal{L}^2$ by passage to the limit. $\square$

Since $|X(t)|$ is a submartingale whenever $X(t)$ is a martingale, Theorems 2.7 and 2.6b can be combined with inequalities (3.6) and (3.7) to obtain the following estimates for stochastic integrals.

THEOREM 2.8   Let $f \in \mathcal{L}^2$ and let $X(t) = \int_a^t f(s)\, dW(s)$. Then, for arbitrary $r > 0$,

$$P\left(\sup_{[a,b]} |X(t)| > r\right) \le E\left(\int_a^b f^2(t)\, dt\right) \Big/ r^2 \tag{3.9}$$

and

$$E\left(\sup_{[a,b]}|X(t)|^2\right) \le 4E\left(\int_a^b f^2(t)\,dt\right). \tag{3.10}$$

The main result of this section presented next indicates how (3.9) can be extended to functions in $\mathcal{L}_s$.

THEOREM 2.9 A (separable) version of the stochastic integral $X(t) = \int_a^t f(s)\,dW(s)$ satisfies the following properties:

(a) $X(t)$ is $\mathcal{A}(t)$-measurable.
(b) $X(t)$ has continuous sample paths w.p. 1.
(c) For arbitrary positive numbers r and N

$$P\left(\sup_{[a,b]}|X(t)| > r\right) \le P\left(\int_a^b f^2(t)\,dt > N\right) + \frac{N}{r^2}. \tag{3.11}$$

Proof: Since $X(t)$ is the limit in probability of a sequence $X_n(t)$ of integrals of step functions property (a) follows from the fact that each of the latter is $\mathcal{A}(t)$-measurable.

Property (b) will now be established first for $f \in \mathcal{L}^2$. It is clear that (b) is satisfied for integrals of step functions since the Wiener process itself has continuous sample paths. For $f \in \mathcal{L}^2$, let $\{f_n\}$ be a sequence of step functions such that $f_n \to f$ in $\mathcal{L}^2$. Then taking $X_n(t) = \int_a^t f_n(s)\,dW(s)$, $X - X_n$ is a martingale and from (3.9) it follows that

$$P\left(\sup_{[a,b]}|X(t) - X_n(t)| > r\right) \le E\left(\int_a^b |f(t) - f_n(t)|^2\,dt\right)\Big/ r^2.$$

Furthermore a sequence of positive numbers $\{r_k\}$ converging to 0, and a subsequence $\{f_{n_k}\}$ can be chosen so that

$$\sum_k E\left(\int_a^b |f(t) - f_{n_k}(t)|^2\,dt\right)\Big/ r_k^2 < \infty.$$

Therefore $\sum_k P(\sup_{[a,b]}|X(t) - X_{n_k}(t)| > r_k) < \infty$. By the Borel-Cantelli lemma (Theorem 1.1) there exists for all $\omega \in \Omega$ outside a set of probability 0, an integer $k_0(\omega)$ such that

$$\sup_{[a,b]}|X(t,\omega) - X_{n_k}(t,\omega)| \le r_k$$

for all $k \ge k_0(\omega)$. Hence $X(t)$ is the uniform limit with probability 1 of continuous functions, and therefore (b) follows.

To establish (b) now for random functions $f \in \mathcal{L}_s$, the truncations f_N for $N > 0$ defined by

$$f_N(t) = \begin{cases} f(t) & \text{if } \int_a^t |f(s)|^2 \, ds \leq N \\ \\ 0 & \text{if } \int_a^t |f(s)|^2 \, ds > N \end{cases}$$

are employed. The process

$$X_N(t) = \int_a^t f_N(s) \, dW(s)$$

is continuous by what was proved in the previous paragraph, and $X(t) = X_N(t)$ for all $t \in [a, b]$ on the set

$$A_N = \left\{ \omega \in \Omega : \int_a^b |f(s)|^2 \, ds \leq N \right\}.$$

Since

$$\int_a^b |f(s)|^2 \, ds < \infty \quad \text{w.p. 1},$$

$$P \left(\bigcup_N A_N \right) = 1.$$

Observing that $\bigcup_N A_N$ is contained in the set of ω for which $X(\cdot, \omega)$ is continuous convinces one that the latter set has probability 1.

Property (c) is also verified by using the truncations $\{f_N\}$ defined above. First, for each $t \in [a, b]$,

$$P(|X(t)| > r) \leq P(|X_N(t)| > r) + P(X(t) \neq X_N(t)). \tag{3.12}$$

Since $\{f_N\} \subseteq \mathcal{L}^2$, (3.9) and Theorem 2.6b can be used to estimate the first probability on the right in (3.12); also,

$$\{X(t) \neq X_N(t)\} = \left\{ \int_a^t |f(s)|^2 \, ds > N \right\} \subseteq \left\{ \int_a^b |f(s)|^2 \, ds > N \right\}$$

so that, one obtains from (3.12)

$$P(|X(t)| > r) \leq \frac{N}{r^2} + P \left(\int_a^b |f(s)|^2 \, ds > N \right)$$

from which the result (c) follows by taking the supremum.   □

2.4   ITO'S FORMULA

If f and g are random functions with $g \in \mathcal{L}_s$, and f satisfying (a) and (b) of the definition of $\mathcal{L}_s$ (Section 2.2) and

$$\int_a^b |f(t)|\, dt < \infty \quad \text{w.p. 1,} \tag{4.1}$$

then the equation

$$X(t) = X(a) + \int_a^t f(s)\, ds + \int_a^t g(s)\, dW(s), \qquad a \le t \le b \tag{4.2}$$

defines a stochastic process with continuous sample paths with probability 1. The first integral in (4.2) is an ordinary integral along paths via (4.1) and the second integral is the Ito integral defined in the last section. In this case, the process $X(t)$ is said to possess the *stochastic differential*

$$dX(t) = f(t)\, dt + g(t)\, dW(t). \tag{4.3}$$

Equation (4.3) characterizes any process whose increments $X(t) - X(s)$, $t > s$, are given by

$$X(t) - X(s) = \int_s^t f(u)\, du + \int_s^t g(u)\, dW(u). \tag{4.4}$$

If f and g are not random, it is easy to see that those increments corresponding to disjoint intervals will be independent and Gaussian with

$$E(X(t) - X(s)) = \int_s^t f(u)\, du$$

and

$$V(X(t) - X(s)) = \int_s^t g^2(u)\, du.$$

Another example arises from noticing, from Section 2.1,

$$W^2(t) - W^2(s) = t - s + \int_s^t 2W(u)\, dW(u).$$

In this case, the process $W^2(t)$ is said to have stochastic differential

$$dW^2(t) = dt + 2W(t)\, dW(t), \tag{4.5}$$

which provides a simple example of the following scheme.

   If $F(t, x)$ is a sufficiently smooth deterministic function defined for all $t \in [a, b]$, $x \in R$, and $X(t)$ is a process with stochastic differential (4.3), then a theorem of K. Ito asserts that $F(t, X(t))$ determines a process with stochastic differential

$$dF(t, X(t)) = \tilde{f}(t, X(t))\, dt + \tilde{g}(t, X(t))\, dW(t) \tag{4.6}$$

with Ito's formula giving analytic expressions for $\tilde{f}$ and $\tilde{g}$ in terms of the partial derivatives of F and the functions f and g in the stochastic differential for X, (4.3). In the example (4.5), $F(t, x) = F(x) = x^2$, and $X(t) = W(t)$ with stochastic differential given by taking $f(t) = 0$ and $g(t) = 1$ in (4.3), which results in $\tilde{f}(x) = 1$ and $\tilde{g}(x) = 2x$. For scalar processes, Ito's result can be stated as follows.

THEOREM 2.10   Suppose $X(t)$ has stochastic differential (4.3). If $F(t, x)$ is a real-valued deterministic function defined for all $t \in [a, b]$, $x \in R$, with continuous partial derivatives $\partial F/\partial t$, $\partial F/\partial x$, and $\partial^2 F/\partial x^2$, then the process $F(t, X(t))$ has a stochastic differential $dF(t, X(t))$ on $[a, b]$ given by (4.6) with

$$\tilde{f}(t, x) = \left[\frac{\partial F}{\partial t} + \frac{\partial F}{\partial x}f + \tfrac{1}{2}\frac{\partial^2 F}{\partial x^2}g^2\right](t, x)$$

$$\tilde{g}(t, x) = \left[\frac{\partial F}{\partial x}g\right](t, x).$$

Proof: Let $Y(t) = F(t, X(t))$. Establishing the theorem consists of showing that, for $t > s$,

$$Y(t) - Y(s) = \int_s^t \left[\frac{\partial F}{\partial t} + \frac{\partial F}{\partial x}f + \frac{1}{2}\frac{\partial^2 F}{\partial x^2}g^2\right](u, X(u))\, du$$

$$+ \int_s^t \left[\frac{\partial F}{\partial x}g\right](u, X(u))\, dW(u). \tag{4.7}$$

The result is proved first for the case of f and g constant. Let $s = t_0 < t_1 < \cdots < t_n = t$, and denote the increment $h(t_{k+1}) - h(t_k)$, for any function h, by Δh_k. Then one can write

$$Y(t) - Y(s) = F(t, X(t)) - F(s, X(s)) = \sum_{k=0}^{n-1} \Delta F_k.$$

From Taylor's theorem, there exist positive constants α_k and β_k not exceeding 1, such that

$$\Delta F_k = \frac{\partial F}{\partial t}(t_k + \alpha_k \Delta t_k, X(t_k)) \Delta t_k + \frac{\partial F}{\partial x}(t_k, X(t_k)) \Delta X_k$$

$$+ \frac{1}{2}\frac{\partial^2 F}{\partial x^2}(t_k, X(t_k) + \beta_k \Delta X_k)(\Delta X_k)^2. \tag{4.8}$$

By the continuity of $\partial F/\partial t$ and $\partial^2 F \partial x^2$, and the w.p. 1 continuity of $X(t)$, one obtains, as $\max \Delta t_k \to 0$,

$$\frac{\partial F}{\partial t}(t_k + \alpha_k \Delta t_k, X(t_k)) \longrightarrow \frac{\partial F}{\partial t}(t_k, X(t_k)) \quad \text{w.p. 1}$$

and $\tag{4.9}$

$$\frac{\partial^2 F}{\partial x^2}(t_k, X(t_k) + \beta_k \Delta X_k) \longrightarrow \frac{\partial^2 F}{\partial x^2}(t_k, X(t_k)) \quad \text{w.p. 1.}$$

With f and g constant, $\Delta X_k = f \Delta t_k + g \Delta W_k$, and so

$$\sum_{k=0}^{n-1}(\Delta X_k)^2 - (g\Delta W_k)^2 = f^2 \sum_{k=0}^{n-1}(\Delta t_k)^2 + 2fg\sum_{k=0}^{n-1} \Delta W_k \Delta t_k. \tag{4.10}$$

Both sums on the right side of (4.10) tend to zero in probability as $\max \Delta t_k \to 0$. Taking (4.8) to (4.10) into account, one obtains

$$Y(t) - Y(s) = \lim_{\max \Delta t_k \to 0} \sum \Delta F_k$$

$$= \lim \sum \left\{ \left[\frac{\partial F}{\partial t} + \frac{\partial F}{\partial x}f + \frac{1}{2}\frac{\partial^2 F}{\partial x^2}g^2 \right](t_k, X(t_k)) \right\} \Delta t_k$$

$$+ \lim \sum \left\{ \left[\frac{\partial F}{\partial x}g \right](t_k, X(t_k)) \right\} \Delta W_k$$

$$+ \lim \sum \left\{ \left[\frac{1}{2}\frac{\partial^2 F}{\partial x^2}g^2 \right](t_k, X(t_k)) \right\}$$

$$\times \left\{ (\Delta W_k)^2 - \Delta t_k \right\} \tag{4.11}$$

where the limits are taken in probability. (Recall that this is the mode of convergence in S.) The first two limits on the right side of (4.11) are the terms on the right side of (4.7). It remains to show that the last limit in (4.11) is zero.

Let $\gamma_k = (\Delta W_k)^2 - \Delta t_k$ and let $I_k^{(N)}$ denote the indicator function of the event

$$\{|X(t_i)| \le N, \quad \text{for all } i \le k\}.$$

Now, since the γ_k are independent, $E\gamma_k = 0$, and $E\gamma_k^2 = 2(\Delta t_k)^2$, it follows that

$$E\left(\sum \frac{\partial^2 F}{\partial x^2}(t_k, X(t_k)) I_k^{(N)} \gamma_k\right)^2 = \sum E\left(\frac{\partial^2 F}{\partial x^2}(t_k, X(t_k)) I_k^{(N)} \gamma_k\right)^2$$

$$\le \left(\sup_{\substack{s<u\le t \\ |x|\le N}}\left[\frac{\partial^2 F}{\partial x^2}(u, x)\right]^2\right) \sum 2(\Delta t_k)^2 \longrightarrow 0$$

$$\text{as } \max \Delta t_k \longrightarrow 0.$$

This establishes that the last limit in (4.11) is zero provided

$$P(|X(t_i)| \le N, \text{ all } i \le k) \longrightarrow 1, \qquad \text{as } N \longrightarrow \infty. \tag{4.12}$$

To see that (4.12) holds let

$$Q_N = \left\{\sup_{s\le u\le t} |X(u)| > N\right\};$$

a straightforward extension of (3.11) implies that $P(Q_N) \to 0$, as $N \to \infty$. Then (4.12) follows, since

$$\left\{I_k^{(N)} = 0, \text{ some } k\right\} = \bigcup_k \{|X(t_i)| > N, \text{ some } i < k\} \subseteq Q_N;$$

the proof is complete, for the case of f and g constant functions.

It is clear that the result follows for the case where f and g are step functions (since on subintervals of an appropriate partition of the interval f and g are constant).

For the general case, suppose $\{f_n(u)\}$ and $\{g_n(u)\}$ are sequences of step functions such that, in probability (mode of convergence in $\mathcal{L}_s$),

$$\int_s^t |f_n(u) - f(u)|\, du \longrightarrow 0 \qquad \text{and} \qquad \int_s^t |g_n(u) - g(u)|^2\, du \longrightarrow 0.$$

Then the sequence

$$X_n(u) = X(s) + \int_s^u f_n(r)\, dr + \int_s^u g_n(r)\, dW(r)$$

converges in probability to $X(u)$ as $n \to \infty$. By considering subsequences, without loss of generality, in each of these cases convergence in probability can be replaced by convergence with probability 1, and furthermore $X_n \to X$ uniformly on $[s,t]$. Ito's formula holding for step functions means that, for each n,

$$Y_n(t) - Y_n(s) = F(t, X_n(t)) - F(s, X_n(s))$$

$$= \int_s^t \left[\frac{\partial F}{\partial t} + \frac{\partial F}{\partial x} f_n + \frac{1}{2} \frac{\partial^2 F}{\partial x^2} g_n^2 \right] (u, X_n(u)) \, du$$

$$+ \int_s^t \left[\frac{\partial F}{\partial x} g_n \right] (u, X_n(u)) \, dW(u). \qquad (4.13)$$

The left-hand side of (4.13) converges with probability 1 to $Y(t) - Y(s)$. Since $X_n \to X$, $f_n \to f$, and $g_n \to g$ in probability, a triangle inequality argument yields, in the stochastic convergence sense,

$$\left[\frac{\partial F}{\partial t} + \frac{\partial F}{\partial x} f_n + \frac{1}{2} \frac{\partial^2 F}{\partial x^2} g_n^2 \right] (u, X_n(u))$$

$$\longrightarrow \left[\frac{\partial F}{\partial t} + \frac{\partial F}{\partial x} f + \frac{1}{2} \frac{\partial^2 F}{\partial x^2} g^2 \right] (u, X(u)) \quad (4.14)$$

and

$$\left[\frac{\partial F}{\partial x} g_n \right] (u, X_n(u)) \longrightarrow \left[\frac{\partial F}{\partial x} g \right] (u, X(u)). \qquad (4.15)$$

In fact, by possibly taking subsequences, (4.14) may be considered to hold with probability 1. Since X is continuous w.p. 1., it is bounded w.p. 1; thus, since all functions appearing in (4.14) are subsequently bounded, the dominated convergence theorem may be applied on each continuous path to obtain that the first integral in (4.13) tends to

$$\int_s^t \left[\frac{\partial F}{\partial t} + \frac{\partial F}{\partial x} f + \frac{1}{2} \frac{\partial^2 F}{\partial x^2} g^2 \right] (u, X(u)) \, du \qquad \text{w.p. 1.}$$

Similarly, from (4.15)

$$\int_s^t \left[\frac{\partial F}{\partial x} g_n \right]^2 (u, X_n(u)) \, du \longrightarrow \int_s^t \left[\frac{\partial F}{\partial x} g \right]^2 (u, X(u)) \, du \qquad \text{w.p. 1,}$$

which implies convergence in the $\mathcal{L}_s$ sense. Consequently, the second integral in (4.13) tends to

$$\int_s^t \left[\frac{\partial F}{\partial x} g \right] (u, X(u)) \, dW(u)$$

in probability, and once again, by possibly considering subsequences, the convergence is w.p. 1. Thus the right-hand side of (4.13) tends to that of (4.7), and the proof for the general case is complete.   □

Note that (4.14) and (4.15) are preserved under multiplication by the indicator function $I_{\{u<\tau\}}$, where τ is a Markov random time satisfying $\tau \leq t$ w.p. 1. The result is that Ito's formula holds for such Markov times τ, that is, (4.7) is valid with t replaced by such a τ.

What is peculiar about Ito's formula is the last term in $\tilde{f}$, that is, $\frac{1}{2}[\partial^2 F/\partial x^2]g^2$. Formally the appearance of this term can be explained as follows. One expects the formula for the differential

$$dF(t,x) = \frac{\partial F}{\partial t}\,dt + \frac{\partial F}{\partial x}\,dx$$

since by Taylor's theorem the higher-order terms

$$\frac{1}{2}\frac{\partial^2 F}{\partial t^2}(dt)^2 + \frac{\partial^2 F}{\partial t\partial x}(dt)(dx) + \frac{1}{2}\frac{\partial^2 F}{\partial x^2}(dx)^2 + \cdots \tag{4.16}$$

are $o(dt)$ or $o(dx)$ as $dt \to 0$ and $dx \to 0$. But

$$dx = f\,dt + g\,dW$$

is not independent of dt, and

$$(dx)^2 = f^2(dt)^2 + 2fg\,dt\,dW + g^2(dW)^2. \tag{4.17}$$

The key point is that $(dW)^2$ behaves like dt in mean square calculus. Although the first two of the higher-order terms in (4.16) are $o(dt)$, and the first two summands of the third term [when (4.17) is substituted into (4.16)] are also $o(dt)$, the last summand $\frac{1}{2}\left[\partial^2 F/\partial x^2\right]g^2(dW)^2$ behaves like

$$\frac{1}{2}\frac{\partial^2 F}{\partial x^2}g^2\,dt$$

and so must be included in the correct expression for the differential. Establishing that the last limit in (4.11) is zero is the rigorous justification of this heuristic argument.

This formal approach is now used to explain Ito's formula for the n-vector $X(t)$ with stochastic differential

$$dX(t) = f(t)\,dt + G(t)\,dW(t). \tag{4.18}$$

Here $W(t) = (W_1(t),\ldots,W_m(t))^T$ is an m-dimensional Wiener process with independent components, $f(t) = (f_1(t),\ldots,f_n(t))^T$ is an n-vector function with L^1 components with probability 1, and $G(t) = \{g_{ij}(t)\}$ is an $n \times m$-matrix function with components g_{ij} in $\mathcal{L}_s$. In component form, (4.18) can be written

$$dx_i = f_i \, dt + \sum_{j=1}^{m} g_{ij} \, dW_j. \qquad (4.18a)$$

Again, one might expect that the differential, for $F(t, x_1, \ldots, x_n)$ a smooth function, would have the form

$$dF = \frac{\partial F}{\partial t} \, dt + \frac{\partial F}{\partial x} \, dx$$

$$= \frac{\partial f}{\partial t} \, dt + \sum_{i=1}^{n} \frac{\partial F}{\partial x_i} \left(f_i \, dt + \sum_{j=1}^{m} g_{ij} \, dW_j \right).$$

But once again, the second-order Taylor terms

$$\frac{1}{2} \frac{\partial^2 F}{\partial x_i \partial x_j} dx_i dx_j$$

include summands that are $o(dt)$. Indeed,

$$dx_i dx_j = f_i f_j (dt)^2 + \sum_{k=1}^{m} (f_i g_{jk}) \, dW_k \, dt$$

$$+ \sum_{k \neq h} g_{ik} g_{jh} \, dW_k \, dW_h + \sum_{k=1}^{m} g_{ik} g_{jk} (dW_k)^2. \qquad (4.19)$$

The first terms in (4.19) are $o(dt)$, and due to the independence of W_k and W_h and the mean square calculus the $dW_k \, dW_h$ terms, for $k \neq h$, vanish. Thus the contribution of each of these second-order terms to the differential is

$$\frac{1}{2} \frac{\partial^2 F}{\partial x_i \partial x_j} \sum_{k=1}^{m} g_{ik} g_{jk} \, dt.$$

Ito's formula, which is established rigorously similarly to the one-dimensional case, then follows

$$dF = \left(\frac{\partial F}{\partial t} + \sum_{i=1}^{n} \frac{\partial F}{\partial x_i} f_i + \sum_{i,j=1}^{n} \sum_{k=1}^{m} \frac{1}{2} \frac{\partial^2 F}{\partial x_i \partial x_j} g_{ik} g_{jk} \right) dt$$

$$+ \sum_{i=1}^{n} \frac{\partial F}{\partial x_i} \sum_{j=1}^{m} g_{ij} \, dW_j. \qquad (4.20)$$

In vector-matrix notation, (4.20) can be written

$$dF = \left[F_t + f^T F_x + \tfrac{1}{2} \operatorname{tr}(GG^T F_{xx}) \right] dt + F_x^T G\, dW. \qquad (4.20a)$$

Here, F_x and F_{xx} denote the gradient of F and the matrix of second partial derivatives of F respectively, and tr means trace or sum of the main diagonal entries.

   EXAMPLE 2.1   Consider the scalar processes $X_i(t)$ having stochastic differentials

$$dX_i(t) = f_i(t)\, dt + g_i(t)\, dW_i(t), \qquad i = 1, 2.$$

If $W_1(t)$ and $W_2(t)$ are independent Wiener processes, then the product $Y(t) = X_1(t) X_2(t)$ has stochastic differential

$$
\begin{aligned}
dY(t) &= X_2(t) dX_1(t) + X_1(t) dX_2(t) \\
&= [X_2(t) f_1(t) + X_1(t) f_2(t)]\, dt \\
&\quad + X_2(t) g_1(t)\, dW_1(t) + X_1(t) g_2(t)\, dW_2(t).
\end{aligned} \qquad (4.21)
$$

If $W_1(t) = W_2(t) = W(t)$, then

$$
\begin{aligned}
dY(t) &= X_2(t) dX_1(t) + X_1(t) dX_2(t) + g_1(t) g_2(t)\, dt \\
&= [X_2(t) f_1(t) + X_1 f_2(t) + g_1(t) g_2(t)]\, dt \\
&\quad + [X_2(t) g_1(t) + X_1(t) g_2(t)]\, dW(t).
\end{aligned} \qquad (4.22)
$$

   EXAMPLE 2.2   Assume $G(t)$ is an $n \times m$ matrix function, $W(t)$ is an m-dimensional Wiener process with independent components, and the process $X(t)$ has stochastic differential

$$dX(t) = G(t)\, dW(t)$$

Then the process $|X(t)|^{2k}$ has stochastic differential

$$
\begin{aligned}
d|X(t)|^{2k} &= 2k|X(t)|^{2k-2} X(t)^T G(t)\, dW(t) \\
&\quad + \big\{ k|X(t)|^{2k-2}|G(t)|^2 + 2k(k-1)|X(t)|^{2k-4} \\
&\qquad \times |X(t)^T G(t)|^2 \big\}\, dt,
\end{aligned} \qquad (4.23)
$$

if k is any positive integer.

   The application of Ito's formula obtaining (4.23) leads to the following estimate for the even order moments of stochastic integral $X(t)$.

THEOREM 2.11   Assume $G(t) = \{g_{ij}(t)\}$ is an $n \times m$ matrix function in $\mathcal{L}_s$ on the interval $[a, b]$. If, for k a positive integer,

$$\int_a^b E|G(t)|^{2k}\, dt < \infty,$$

then

$$E\left| \int_a^b G(t)\, dW(t) \right|^{2k} \leq [k(2k-1)]^k (b-a)^{k-1} \int_a^b E|G(t)|^{2k}\, dt. \quad (4.24)$$

Proof: From Example 2.2, the integrated version of (4.23) is

$$\left| \int_a^t G(s)\, dW(s) \right|^{2k}$$

$$= 2k \int_a^t \left| \int_a^s G(r)\, dW(r) \right|^{2k-2} \left[\int_a^s G(r)\, dW(r) \right]^T G(s)\, dW(s)$$

$$+ \int_a^t \left\{ k \left| \int_a^s G(r)\, dW(r) \right|^{2k-2} \right. \quad (4.25)$$

$$\times\, |G(s)|^2 + 2k(k-1) \left| \int_a^s G(r)\, dW(r) \right|^{2k-4}$$

$$\left. \times \left| \left[\int_a^s G(r)\, dW(r) \right]^T G(s) \right|^2 \right\} ds.$$

Without loss of generality (G can be approximated by such functions) assume G is bounded. Then the integrand in the first integral on the right in (4.25) is in $\mathcal{L}^2$ and so has mean zero. Taking expected value in (4.25) (and noticing that the two terms in the brackets above are the same except for the constant factors k and $2k(k-1)$, respectively, one has, for $t = b$

$$E\left| \int_a^b G(s)\, dW(s) \right|^{2k}$$

$$\leq k(2k-1) \int_a^b E\left(\left| \int_a^t G(s)\, dW(s) \right|^{2k-2} |G(t)|^2 \right) dt. \quad (4.26)$$

By Hölder's inequality applied to the right side of (4.26)

$$
E\left|\int_a^b G(s)\,dW(s)\right|^{2k} \leq k(2k-1)
$$

$$
\times\left\{\int_a^b E\left|\int_a^t G(s)\,dW(s)\right|^{2k}dt\right\}^{(2k-2)/2k}
$$

$$
\times\left\{\int_a^b E|G(t)|^{2k}\,dt\right\}^{2/2k} \tag{4.27}
$$

Taking expected value in (4.25) indicates that

$$
E\left|\int_a^t G(s)\,dW(s)\right|^{2k}
$$

is a nondecreasing function of t, so (4.27) can be written

$$
E\left|\int_a^b G(s)\,dW(s)\right|^{2k} \leq k(2k-1)
$$

$$
\times\left\{E\left|\int_a^b G(s)\,dW(s)\right|^{2k}(b-a)\right\}^{1-1/k}
$$

$$
\times\left\{\int_a^b E|G(t)|^{2k}\,dt\right\}^{1/k}. \tag{4.28}
$$

Now, raising both sides of (4.28) to the kth power and dividing both sides by $\left\{E\left|\int_a^b G(s)\,dW(s)\right|^{2k}\right\}^{k-1}$ yields the result.   □

EXERCISES

2.1   Prove Theorem 2.4 by computing the moments directly from the value of the integral

$$
\int_a^b W(t)\,dW(t) = \tfrac{1}{2}\left\{W^2(b) - W^2(a) - (b-a)\right\}.
$$

2.2   Verify formula (1.19).

2.3   Show that convergence in S with respect to the $\|\ \|_S$ metric is equivalent to convergence in probability.

2.4   Prove that $\mathcal{E}$ is dense in $\mathcal{L}_s$.

2.5   Verify, using Theorem 2.8, that

(a) $P\left(\sup_{[a,b]} |W(t)| > r\right) \le (b - a)/r^2$.
(b) $E\left(\sup_{[a,b]} |W(t)|^2\right) \le 4(b - a)$.

2.6   Use Ito's formula to verify that

$$\int_a^b W^n(t)\, dW(t) = \frac{1}{n+1}[W^{n+1}(b) - W^{n+1}(a)]$$

$$- \frac{n}{2} \int_a^b W^{n+1}(t)\, dt.$$

for all positive integers n.

2.7   Use Theorem 2.11 to obtain an estimate for $E|W(b) - W(a)|^4$. Compare with (9.1) of Chapter 1.

3

Basic Theory of Stochastic
Differential Equations

3.0 ORDINARY, RANDOM, AND STOCHASTIC DIFFERENTIAL
EQUATIONS: FUNDAMENTAL QUESTIONS AND PROPERTIES

Ordinary differential equations which have the general form

$$\frac{dx}{dt} = f(t, x) \tag{0.1}$$

provide simple deterministic descriptions of the laws of motion of physical systems. The solution $x(t)$ of an initial value problem consisting of (0.1) together with an initial value

$$x(t_0) = x_0 \tag{0.2}$$

represents the state of such a system at time $t > t_0$, given that the state (0.2) was attained at time t_0. If random aspects in the physical system are to be considered, a number of modifications can be made in the formulation of the initial value problem (0.1), (0.2). The initial point x_0 may be replaced by a random variable X_0; the deterministic function $f(t, x)$ may be replaced by a random function $F(t, X, Y)$, where $Y = Y(t)$ designates a random input process uncoupled with the solution variable X; or, in the latter case, Y may represent the random coefficients of a linear or nonlinear operator whose form is specified by F. The first of these three possibilities for randomizing (0.1), (0.2) is exemplified by the motion of a space vehicle whose state consisting of position and momentum components changes according to a deterministic law but whose initial values may be subject to some uncertainty. An ac electric power circuit with state described by voltages and phase angles at nodes whose rates of change are forced by noise is a particular case of the second type of randomness. An example of the third type arises by considering intrinsic birth-death rates and interaction rates as stochastic processes in differential equation models of multispecies population evolution; a random initial value problem where the stochastic input process is coupled with the solution results. In recent literature [see Soong (1973), for example] the term *random differential equation* is reserved for the last of these three types, while *stochastic differential equation* refers to equations of the second type, which are driven by white noise and interpreted mathematically as Ito equations.

The ultimate goal of the analysis of any random initial value problem

$$\frac{dX(t)}{dt} = F(t, X(t), Y(t)) \tag{0.3}$$

$$X(t_0) = X_0, \tag{0.4}$$

generally, is to obtain the distribution of the solution process $X(t)$ in terms of the distributions of X_0, $Y(t)$, and the statistical and deterministic properties of F; determining the sample path structure of $X(t)$ is even more ambitious. When randomness enters the problem only through the initial condition, a situation sometimes called "cryptodeterministic," the solution is a deterministic transformation of the random variable X_0. The mean square theory for the random initial value problem, in this instance, is a direct analogue of the ordinary differential equations theory [see Soong (1973), chapter 6]. For ordinary differential equations, closed-form expressions for solutions are often unobtainable, and so one must be satisfied with numerical approximations or less than complete qualitative information about solutions. It is unreasonable to expect otherwise for random or stochastic differential equations. However, additional difficulties, related particularly to the stochastic nature of the equations, appear when the equation contains random coefficients, that is, in the random differential equations case. The following example illustrates this situation. Consider the initial value problem (0.3), (0.4) for the scalar process $X(t)$ where

$$F(t, X(t), Y(t)) = Y_1(t)X(t) + Y_2(t) \tag{0.5}$$

and $Y(t) = (Y_1(t), Y_2(t))$ is a two-dimensional random process. The formal solution of (0.3), (0.4), (0.5)

$$X(t) = X_0 \exp\left[\int_{t_0}^t Y_1(s)\, ds\right] + \int_{t_0}^t Y_2(u) \exp\left[\int_u^t Y_1(s)\, ds\right] du$$

is valid only if Y_1 and $Y_2 \exp \int Y_1$ are mean square or with probability 1 integrable; determining the distribution of $X(t)$ requires a priori knowledge of the joint distribution of X_0, Y_1, and Y_2. Suppose, on the other hand, one writes the initial value problem as the integral equation

$$X(t) = X_0 + \int_{t_0}^{t} Y_1(s)X(s)\,ds + \int_{t_0}^{t} Y_2(s)\,ds. \qquad (0.6)$$

From (0.6) one obtains an equation for the mean of the solution

$$EX(t) = EX_0 + \int_{t_0}^{t} E(Y_1(s)X(s))\,ds + \int_{t_0}^{t} EY_2(s)\,ds \qquad (0.7)$$

[provided taking expected value commutes with the sample path integration on the right-hand side of (0.6)]. Equation (0.7), however, is difficult to analyze unless the correlation of Y_1 and X is known. In particular, the mean of the solution does *not* satisfy, in general, the averaged differential equation

$$\frac{dx}{dt} = E(Y_1(t))x + E(Y_2(t)),$$

even in this, the linear case. [Soong (1973, chap. 8) gives a method, involving characteristic functions, for computing the mean and correlation of the solution of this problem assuming $X_0 = 0$ and that Y_1 and Y_2 are stationary and correlated Gaussian processes with known means and correlation functions.] Although equations with random coefficients are perhaps the most interesting from the point of view of applications as the most general random analogue of ordinary differential equations, this simple example indicates their intractability in general, even in the linear case. Specific a priori information concerning the correlation of the random components of the equation is required to make progress in a mathematical analysis of the solution.

This monograph is concerned with Eq. (0.3) where

$$F(t, X(t), Y(t)) = f(t, X(t)) + g(t, X(t))Y(t), \qquad (0.8)$$

with $Y(t)$ representing a Gaussian white noise process. As illustrated in Chapter 2, the problem of defining the stochastic integral in the corresponding integral equation indicates that such an equation is at best mathematically ambiguous. More precisely, then, of interest here is the specific interpretation of (0.3), (0.8) as the Ito equation

$$dX(t) = f(t, X(t))\,dt + g(t, X(t))\,dW(t) \qquad (0.9)$$

where $W(t)$ denotes a Wiener or Brownian motion process; in (0.9), f and g are deterministic functions, but with slight modifications, the theory extends to explicitly random functions. Stochastic differential equations of this type were introduced by K. Ito in 1942, and the basic theory was developed independently by Ito and I. Gihman during the 1940s. Applications to control problems in electrical engineering motivated by the need for more sophisticated models spurred further work on these equations in the 1950s and 1960s. In the last decade applications have been extended to other areas including population dynamics in biology. The principal motivation for choosing the Ito approach (as opposed to Stratonovich calculus, the other most popular interpretation of (0.8)) is that the Ito method extends to a broader class of equations and transforms the probability law of the Wiener process in a more natural way; from the applications point of view, the Ito interpretation implements a diffusion approximation that arises from random difference equation models, a situation that will be discussed in Chapter 6.

The basic theory on existence and uniqueness of solutions and dependence of solutions on initial conditions and parameters presented in the next section emphasizes the mathemtical tractability of this class of equations from the differential equations viewpoint. On the other hand, the third section of this chapter deals with the correspondence between solutions and Markov diffusion processes, an important class of stochastic processes which has an extensively developed theory; thus this section's material also motivates stochastic differential equations but from a probability theory standpoint. Combining both aspects, Ito's formula for solutions of stochastic differential equations, together with an application which yields estimates for the moments of solutions, is given in Section 3.2. The Stratonovich-interpreted stochastic differential equations are discussed in Section 3.4. The last section illustrates how Ito's formula may be applied to characterize important functionals of the solution process, such as exit time statistics, as solutions of initial boundary value problems involving the associated Ito partial differential operator.

The principal results in this chapter, for simplicity, are formulated for scalar stochastic differential equations (0.9), that is, the functions f and g and the processes X and W are all assumed to be real-valued. It is clear that, with certain obvious modifications, these results extend to the vector equation (or system of scalar equations)

$$dX(t) = f(t, X(t))\, dt + G(t, X(t))\, dW(t) \tag{0.10}$$

where f is an n-vector-valued function, G is an $n \times m$-matrix-valued function, $W(t)$ is an m-dimensional process having independent scalar Wiener process components, and the solution $X(t)$ is an n-vector process. Some remarks concerning the extensions to the vector case are included in Sections 3.2, 3.3, and 3.4.

Finally, also for simplicity, $t_0 = 0$ will be assumed throughout this chapter without loss of generality.

3.1 EXISTENCE AND UNIQUENESS OF SOLUTIONS; DEPENDENCE ON PARAMETERS AND INITIAL CONDITIONS

By a *solution* $X(t)$ of the stochastic differential equation (0.9) is meant a process $X(t)$ which has stochastic differential (0.9) as defined in Section 2.4. Equivalently, for all t in some interval $[0, T]$, $X(t)$ must satisfy the integral equation

$$X(t) = X_0 + \int_0^t f(s, X(s))\, ds + \int_0^t g(s, X(s))\, dW(s) \tag{1.1}$$

where X_0 is a specified initial value $X(0)$, the first integral in (1.1) is an ordinary one along paths, and the second integral in (1.1) is the Ito stochastic integral. In this section the simplest existence and uniqueness theorem for such solutions is presented.

First of all, for the right side of (1.1) to be defined in this sense, the random functions $|f(\cdot, X(\cdot))|^{1/2}$ and $g(\cdot, X(\cdot))$ must be in the space $\mathcal{L}_s$ (taking the interval $[a, b] = [0, T]$) defined in Section 2.2, with the underlying family of σ-algebras $\mathcal{A}(t)$ generated by X_0 and $W(s)$, $s \leq t$. If $f(t, x)$ and $g(t, x)$ are bounded measurable (in Lebesgue sense) real-valued functions for all t and x, then it is not difficult to see that the conditions on the corresponding random functions will be satisfied. Boundedness of f and g is not necessary, however. In fact, the following basic theorem verifies existence and uniqueness of solutions without the boundedness assumption provided certain additional conditions concerning the dependence of f and g on the variable x are imposed; in particular these conditions require f and g to be Lipschitz continuous and exhibit linear growth in x.

THEOREM 3.1   Suppose that:

1. The functions $f(t, x)$ and $g(t, x)$ are measurable with respect to t and x, for $t \in [0, T]$ and $x \in R$.
2. There exists a constant K such that for all $t \in [0, T]$, and $x, y \in R$
    (a) $|f(t, x) - f(t, y)| + |g(t, x) - g(t, y)| \leq K|x - y|$.
    (b) $|f(t, x)|^2 + |g(t, x)|^2 \leq K^2(1 + |x|^2)$.
3. X_0 is independent of $W(t)$, for $t > 0$, and $EX_0^2 < \infty$.

Then there is a solution $X(t)$ of (1.1) defined on $[0, T]$ which is continuous with probability 1, and such that

$$\sup_{[0,T]} EX^2(t) < \infty.$$

Furthermore a solution with these properties is *pathwise unique*, that is, if X and Y are two such solutions

$$P\left(\sup_{[0,T]} |X(t) - Y(t)| = 0\right) = 1.$$

EXAMPLE 3.1 The choices

$$f(t, x) = f_1(t) + f_2(t)x \qquad \text{and} \qquad g(t, x) = g_1(t) + g_2(t)x$$

where the functions f_i and g_i are continuous on $[0, T]$, satisfy conditions 1 and 2 in Theorem 3.1. Hence the result applies to the stochastic equation

$$dX(t) = [f_1(t) + f_2(t)X(t)] \, dt + [g_1(t) + g_2(t)X(t)] \, dW(t).$$

The following parts of the proof are separated out as lemmas.

LEMMA 3.2 (BELLMAN-GRONWALL INEQUALITY) If $a(t)$ and $b(t)$ are measurable bounded functions such that for some $L > 0$,

$$a(t) \le b(t) + L \int_0^t a(s) \, ds,$$

then

$$a(t) \le b(t) + L \int_0^t e^{L(t-s)} b(s) \, ds. \tag{1.2}$$

LEMMA 3.3 Let f and g satisfy conditions 1 and 2 in Theorem 3.1. Suppose for $i = 1, 2$, $Y_i(t)$ is a stochastic process defined on $[0, T]$ such that

1. The σ-algebra generated by $Y_i(s)$ and $W(s)$, $s \le t$, is independent of the increments $W(t + r) - W(t)$, $r > 0$.
2. $\sup_{[0,T]} EY_i^2(t) < \infty$.

If the process Z_i is defined on $[0, T]$ by

$$Z_i(t) = X_0 + \int_0^t f(s, Y_i(s)) \, ds + \int_0^t g(s, Y_i(s)) \, dW(s), \tag{1.3}$$

then there is a constant $L > 0$ such that

$$E[Z_1(t) - Z_2(t)]^2 \le L \int_0^t E[Y_1(s) - Y_2(s)]^2 \, ds. \tag{1.4}$$

Proofs: The proof of Lemma 3.2 can be found in Gihman and Skorohod (1972, p. 41) and here is left as an exercise (see Exercise 3.2). To verify Lemma 3.3, one starts with the estimate

$$[Z_1(t) - Z_2(t)]^2 \le 2 \left[\int_0^t [f(s, Y_1(s)) - f(s, Y_2(s))] \, ds \right]^2$$

$$+ 2 \left[\int_0^t [g(s, Y_1(s)) - g(s, Y_2(s))] \, dW(s) \right]^2. \quad (1.5)$$

Using the Cauchy-Schwarz inequality argument

$$\left[\int_0^t h(s) \, ds \right]^2 \le \int_0^t 1^2 \, ds \int_0^t h^2(s) \, ds = t \int_0^t h^2(s) \, ds$$

on the first term on the right side of (1.5), together with assumption 2a of Theorem 3.1 on f, one obtains

$$\left[\int_0^t f(s, Y_1(s)) - f(s, Y_2(s)) \, ds \right]^2 \le K^2 t \int_0^t [Y_1(s) - Y_2(s)]^2 \, ds.$$

$$(1.6)$$

Since $E \int_0^t [g(s, Y_1(s)) - g(s, Y_2(s))]^2 \, ds \le K^2 E \int_0^t [Y_1(s) - Y_2(s)]^2 \, ds < \infty$ by (2), then $g(\cdot, Y_1(\cdot)) - g(\cdot, Y_2(\cdot)) \in \mathcal{L}^2$. Therefore

$$E \left[\int_0^t [g(s, Y_1(s)) - g(s, Y_2(s))] \, dW(s) \right]^2$$

$$= E \int_0^t [g(s, Y_1(s)) - g(s, Y_2(s))]^2 \, ds. \quad (1.7)$$

The lemma follows now by taking expected value in (1.5), using (1.6) and (1.7), and setting

$$L = 2(T + 1)K^2.$$

Proof of Theorem 3.1:   As in the case of ordinary differential equations, the sequence of successive approximations $\{X_n\}$ defined by

$$X_0(t) = X_0$$

$$X_n(t) = X_0 + \int_0^t f(s, X_{n-1}(s)) \, ds + \int_0^t g(s, X_{n-1}(s)) \, dW(s) \qquad (1.8)$$

is shown to converge to the unique solution $X(t)$. By the assumption of this theorem, and induction, (1.8) defines a sequence of processes with continuous sample paths w.p. 1 (Theorem 2.8). The proof can be broken down into separate steps verifying that

(i)   The sequence $\{X_n\}$ is uniformly mean-square bounded on $[0, T]$.
(ii) $\{X_n\}$ is uniformly mean square convergent.
(iii) $\{X_n\}$ is uniformly convergent w.p. 1.
(iv) $\text{limit}_{n \to \infty} X_n$ is a solution of (1.1).
(v)   Any two solutions X and Y of (1.1) agree w.p. 1.

To begin, then, it is shown that the sequence is uniformly bounded; that is, for some constant M and all positive integers n,

$$\sup_{[0,T]} EX_n^2(t) \leq M < \infty. \tag{1.9}$$

Toward verifying (1.9), making use of the inequality $(a + b + c)^2 \leq 3(a^2 + b^2 + c^2)$, one has

$$EX_n^2(t) \leq 3 \left\{ EX_0^2 + E \left[\int_0^t f(s, X_{n-1}(s)\, ds \right]^2 \right.$$

$$\left. + E \left[\int_0^t g(s, X_{n-1}(s))\, dW(s) \right]^2 \right\}.$$

From the inductive hypothesis $\sup_{[0,T]} EX_{n-1}^2(t) < \infty$, and estimates similar to those used in the proof of Lemma 3.3, one obtains

$$EX_n^2(t) \leq 3 \left\{ EX_0^2 + TE \int_0^t f^2(s, X_{n-1}(s))\, ds \right.$$

$$\left. + E \int_0^t g^2(s, X_{n-1}(s))\, ds \right\}.$$

Taking into consideration condition 2b, and interchanging path integration and expected value, this estimate becomes

$$EX_n^2(t) \leq 3 \left\{ EX_0^2 + TK^2 \int_0^t [1 + EX_{n-1}^2(s)]\, ds \right.$$

$$\left. + K^2 \int_0^t [1 + EX_{n-1}^2(s)]\, ds \right\}$$

$$\leq 3 \left\{ EX_0^2 + L \int_0^t [1 + EX_{n-1}^2(s)]\, ds \right\}$$

$$= 3 \left\{ EX_0^2 + Lt + L \int_0^t EX_{n-1}^2(s)\, ds \right\}.$$

Iterating this last inequality results in

$$EX_n^2(t) \le 3 \Big\{ EX_0^2 + Lt$$

$$+ L \int_0^t 3 \Big(EX_0^2 + Ls + L \int_0^s EX_{n-2}^2(r)\, dr \Big)\, ds \Big\}$$

$$= 3 \Big\{ (1 + 3Lt) EX_0^2 + Lt + 3L^2 \frac{t^2}{2} L^2$$

$$+ 3L^2 \int_0^t (t-s) EX_{n-2}^2(s)\, ds \Big\}$$

Continuing, $n - 1$ iterations lead to

$$EX_n^2(t) \le 3 \Big\{ \Big(1 + 3Lt + \cdots + \frac{1}{(n-1)!}(3Lt)^{n-1} \Big) EX_0^2 \Big\}$$

$$+ 3Lt + \cdots + \frac{1}{n!}(3Lt)^n \tag{1.10}$$

$$+ (3L)^n \int_0^t \frac{(t-s)^{n-1}}{(n-1)!} EX_0^2\, ds.$$

Here the equality

$$\int_0^t (t-s)^n \int_0^s h(r)\, dr\, ds = \int_0^t \frac{(t-s)^{n+1}}{n+1} h(s)\, ds \tag{1.11}$$

has been employed. From (1.10) it follows that

$$\sup_{[0,T]} EX_n^2(t) \le (3EX_0^2 + 1)e^{3LT}. \tag{1.12}$$

Observing that the right side of (1.12) is constant establishes (1.9). Hence step (i) is verified—namely, that the sequence of successive approximations $\{X_n\}$ is uniformly bounded.

Toward proving step (ii), the next step in establishing the existence of a solution of equation (1.1), Lemma 3.3 is applied to the sequence of successive approximations defined by equation (1.8) to obtain

$$E[X_{n+1}(t) - X_n(t)]^2 \le L \int_0^t E[X_n(s) - X_{n-1}(s)]^2\, ds. \tag{1.13}$$

Iterating (1.13) and making use of (1.11), one arrives at

$$E[X_{n+1}(t) - X_n(t)]^2 \le L^n \int_0^t \frac{(t-s)^{n-1}}{(n-1)!} E[X_1(s) - X_0(s)]^2\, ds. \tag{1.14}$$

Now proceeding directly from the $n = 1$ case of (1.8), making use of assumption 2b, and the type of estimates used in the previous paragraph, one has

$$E[X_1(t) - X_0(t)]^2 \leq L \int_0^t [1 + EX_0^2] \, ds$$

$$\leq LT[1 + EX_0^2] = C.$$

From (1.14) then it follows that, for all integers $n \geq 0$,

$$\sup_{[0,T]} E[X_{n+1}(t) - X_n(t)]^2 \leq C(LT)^n/n!. \tag{1.15}$$

Inequality (1.15) establishes uniform mean square convergence of the sequence $\{X_n\}$, and so step ii is verified.

But the goal is to prove w.p. 1 uniform convergence of $\{X_n\}$; (1.12) will be of use here. Let

$$Y_n = \sup_{[0,T]} |X_{n+1}(t) - X_n(t)|;$$

$$Y_n \leq \int_0^T |f(s, X_n(s)) - f(s, X_{n-1}(s))| \, ds \tag{1.16}$$

$$+ \sup_{[0,T]} \left| \int_0^t [g(s, X_n(s)) - g(s, X_{n-1}(s))] \, dW(s) \right|.$$

Therefore, by the Cauchy-Schwarz inequality (as used to establish (1.4)) applied to the first term of the right side of (1.16) and by Theorem 2.7 applied to the second term, after squaring, using assumption 2a, and taking expected values, one has

$$EY_n^2 \leq 2K^2 T \int_0^T E[X_n(s) - X_{n-1}(s)]^2 \, ds$$

$$+ 8K^2 \int_0^T E[X_n(s) - X_{n-1}(s)]^2 \, ds \tag{1.17}$$

$$\leq C_1 (LT)^{n-1}/(n-1)!$$

where $C_1 = [2T + 8]CK^2$, and (1.15) has to be employed to establish the last estimate. From (3.9) in Theorem 2.7 and (1.17),

$$P\left\{ Y_n > \frac{1}{n^2} \right\} \leq \frac{C_1 (LT)^{n-1}}{(n-1)!} n^4.$$

The series $\sum_{n=1}^{\infty} P\{Y_n > 1/n^2\}$ is majorized by

$$\sum_{n=1}^{\infty} \frac{C_1(LT)^{n-1}}{(n-1)!} n^4$$

whose convergence implies by the Borel-Cantelli lemma (Theorem 1.1) that

$$P\left\{ \sup_{[0,T]} |X_{n+1}(t) - X_n(t)| \leq \frac{1}{n^2}, \text{ for sufficiently large } n \right\} = 1.$$

Hence, the w.p. 1 uniform convergence of

$$X_n(t) = X_0 + \sum_{k=1}^{n} [X_k(t) - X_{k-1}(t)]$$

to   (1.18)

$$X(t) = X_0 + \sum_{n=1}^{\infty} [X_n(t) - X_{n-1}(t)]$$

is assured.

To see that step (iv) holds, note that $X(t)$ defined by (1.18) is measurable with respect to $A(t)$, the σ-algebra generated by X_0 and $W(s)$, $s \leq t$, since it is the limit of such processes. $X(t)$ is continuous w.p. 1, similarly, as this process is the uniform limit of w.p. 1 continuous processes. That $X(t)$ is mean square bounded on $[0,T]$ follows from (1.9). Taking the limit as $n \to \infty$, in (1.8), one sees that step (iv) holds—namely, that $X(t)$ is a solution of (1.1): indeed, the uniform convergence of $\{X_n\}$ and the Lipschitz condition (2a) imply that

$$\left| \int_0^t f(s, X_n(s))\, ds - \int_0^t f(s, X(s))\, ds \right| \leq K \int_0^t |X_n(s) - X(s)|\, ds,$$

so that the left side tends to zero w.p. 1; furthermore

$$\int_0^t |g(s, X_n(s)) - g(s, X(s))|^2\, ds \leq K^2 \int_0^t |X_n(s) - X(s)|^2\, ds,$$

and so the left side tending to zero w.p. 1 implies that $g(\cdot, X_n(\cdot)) \to g(\cdot, X(\cdot))$ in $\mathcal{L}_s$, which means that

$$\int_0^t g(s, X_n(s))\, dW(s) \longrightarrow \int_0^t g(s, X(s))\, dW(s)$$

in S. (See Section 2.2.)

Finally, uniqueness (v) is established. Suppose X and Y are both solutions to (1.1), which are continuous with probability 1 and mean square bounded on $[0, T]$. The function $E[X(t) - Y(t)]^2$ is a bounded measurable function, and must satisfy, according to Lemma 3.3,

$$E[X(t) - Y(t)]^2 \leq L \int_0^t E[X(s) - Y(s)]^2 \, ds. \tag{1.19}$$

By Lemma 3.2, $E[X(t) - Y(t)]^2 = 0$. Thus, for any $t \in [0, T]$, $P(X(t) = Y(t)) = 1$, and so for any countable subset $B \subseteq [0, T]$

$$P(X(t) = Y(t), t \in B) = 1.$$

Taking B to be dense and making use of w.p. 1 continuity of X and Y,

$$P\left(\sup_{[0,T]} |X(t) - Y(t)| > 0 \right) = 0. \tag{1.20}$$

Therefore (1.20) obtains uniqueness:

$$X(\cdot) = Y(\cdot) \qquad \text{w.p. 1.}$$

This completes the proof of Theorem 3.1. □

Assumption 2 in Theorem 3.1, which essentially requires that the functions f and g satisfy a Lipschitz condition and exhibit linear growth in the state variable, is fairly restrictive. This condition facilitated an elementary proof of the existence and uniqueness of solutions of (1.1), which is analogous to the classical Picard iteration proof given in the ordinary differential equations case; the Borel-Cantelli Lemma is the probabilistic tool employed to deal with the random aspect of the equation. Note that if the Lipschitz condition 2a holds for a function f and if either f is independent of t or for some t, f is bounded in x, then f will satisfy the growth condition 2b (Exercise 3.3). Also, condition 2a can be relaxed to the local Lipschitz condition:

2a'. For each $N > 0$, there is a constant K_N such that for all $t \in [0, T]$, and x and y, $|x| \leq N$, $|y| \leq N$,

$$|f(t, x) - f(t, y)| + |g(t, x) - g(t, y)| \leq K_N |x - y|.$$

(Note that this condition holds whenever f and g are continuously differentiable in the second variable.)

THEOREM 3.4   The conclusion of Theorem 3.1 holds if the hypotheses other than 2a are satisfied, and 2a′ holds.

The proof involves the truncation procedure illustrated in the proof of uniqueness in Theorem 3.1. Details can be found in Gihman and Skorohod (1972).

The purpose of the growth condition 2b, ostensibly, is to guarantee existence of the solution on the entire interval $[0, T]$. However, Protter (1977) has shown recently that the situation here also is entirely analogous to the ordinary differential equations case; that is, Theorem 3.1 is valid even if hypothesis 2b is removed; the Lipschitz condition 2a suffices to guarantee uniqueness and global existence. His proof uses an additional probabilistic tool, namely optional stopping times. The local Lipschitz condition, by itself, does not suffice to give global existence, however. So condition 2b cannot be removed from Theorem 3.4. Indeed, the ordinary initial value problem

$$dX = X^2 \, dt$$
$$X(0) = x_0 \tag{1.21}$$

has as solution

$$X(t) = \begin{cases} 0, & \text{if } x_0 = 0 \\ (1/x_0 - t)^{-1}, & \text{if } x_0 \neq 0, \end{cases}$$

which exhibits "explosions" at the values $1/x_0$. Considering (1.21) as a degenerate stochastic differential equation $[f(t,x) = x^2$ and $g(t,x) = 0]$, one observes that the conditions of Theorem 3.4 except for the growth condition, are satisfied on every interval $[0, T]$, and yet existence cannot be ascertained on any such interval [independent of $X(0)$].

Uniqueness and local existence have been established under weaker conditions than those cited above. Skorohod (1965) has demonstrated existence if f and g are continuous in both variables. Watanabe and Yamada (1971) and Yamada and Watanabe (1971) have given uniqueness results which involve weaker moduli of continuity conditions on f and g in the second variable than the Lipschitz condition. Girsanov's example

$$X(t) = \int_0^t |X(s)|^\alpha \, dW(s),$$

which has one solution for $\alpha \geq 1/2$ but infinitely many for $0 < \alpha < 1/2$, illustrates that some such condition is necessary for uniqueness. Also, this example illustrates that sharper uniqueness results will require different moduli of continuity conditions on f and g since, for any α, $0 < \alpha < 1$, the equation

$$X(t) = \int_0^t |X(s)|^\alpha \, ds$$

has the family of solutions, for $\tau > 0$,

$$X_\tau(t) = \begin{cases} 0, & 0 \le t \le \tau \\ [(1-\alpha)(t-\tau)]^{(1-\alpha)^{-1}}, & t > \tau. \end{cases}$$

When the qualitative theory of stochastic differential equations is discussed, in general, in the next chapter some other uniqueness results will be mentioned.

That solutions of differential equations respond smoothly to smooth changes in the coefficient functions and initial data is important for applications. The corresponding results for stochastic differential equations presented in this section are given in Gihman and Skorohod (1972). As an example the following basic result is included.

THEOREM 3.5   Suppose f_n, g_n, and $X_n(0)$ satisfy the conditions of Theorem 3.1 uniformly in n (same K). Let $X_n(t)$ be the solution of the equation, for $n = 0, 1, 2, \ldots$,

$$X_n(t) = X_n(0) + \int_0^t f_n(s, X_n(s)) \, ds + \int_0^t g_n(s, X_n(s)) \, dW(s).$$

$$(1.22)$$

If

(1) $\displaystyle \lim_{n \to \infty} \sup_{|x| \le N} \left(|f_n(t,x) - f_0(t,x)| + |g_n(t,x) - g_0(t,x)| \right) = 0$

for each $N > 0$ and each $t \in [0, T]$, and

(2) $\displaystyle \lim_{n \to \infty} E[X_n(0) - X_0(0)]^2 = 0$

then

$$\lim_{n \to \infty} \sup_{[0,T]} E[X_n(t) - X_0(t)]^2 = 0.$$

Proof:

$$X_n(t) - X_0(t) = Y_n(t) + \int_0^t [f_n(s, X_n(s)) - f_n(s, X_0(s))] \, ds$$

$$+ \int_0^t [g_n(s, X_n(s)) - g_n(s, X_0(s))] \, dW(s),$$

where

$$Y_n(t) = X_n(0) - X_0(0) + \int_0^t [f_n(s, X_0(s)) - f_0(s, X_0(s))]\, ds$$

$$+ \int_0^t [g_n(s, X_0(s)) - g_0(s, X_0(s))]\, dW(s).$$

Using estimates of the type employed in the proof of Theorem 3.1, replacing the role of the growth condition by the Lipschitz condition, one obtains

$$E[X_n(t) - X_0(t)]^2 \le 3EY_n^2(t) + L \int_0^t E[X_n(s) - X_0(s)]^2\, ds$$

where $L = 3(T+1)K^2$. By Lemma 3.2,

$$E[X_n(t) - X_0(t)]^2 \le 3EY_n^2(t) + L \int_0^t e^{L(t-s)} EY_n^2(s)\, ds$$

and so it suffices to show that

$$\sup_{[0,T]} EY_n^2(t) \longrightarrow 0 \qquad \text{as } n \longrightarrow \infty. \tag{1.23}$$

Toward establishing (1.23), Cauchy-Schwarz can be applied:

$$E\left[\int_0^t f_n(s, X_0(s)) - f_0(s, X_0(s))\, ds\right]^2$$

$$\le TE \int_0^T [f_n(s, X_0(s)) - f_0(s, X_0(s))]^2\, ds. \tag{1.24}$$

The integrand on the right-hand side of (1.24) is dominated by $2K^2(1 + X_0^2(s))$, and $E\int_0^T [1 + X_0^2(s)]\, ds < \infty$. By Lebesgue's dominated convergence theorem the right side of (1.24) tends to zero, and so as $n \to \infty$,

$$\sup_{[0,T]} E\left[\int_0^t f_n(s, X_0(s)) - f_0(s, X_0(s))\, ds\right]^2 \longrightarrow 0, \tag{1.25}$$

Since the function $h_n(\cdot) = g_n(\cdot, X_0(\cdot)) - g_0(\cdot, X_0(\cdot)) \in \mathcal{L}^2$ also, Theorem 2.7 can be applied to obtain

$$E \sup_{[0,T]} \left[\int_0^t g_n(s, X_0(s)) - g_0(s, X_0(s))\, dW(s)\right]^2$$

$$\le 4E \int_0^T [g_n(s, X_0(s)) - g_0(s, X_0(s))]^2\, ds. \tag{1.26}$$

Again the right side of (1.26) $\to 0$ as $n \to \infty$. Together with (1.25) and assumption 2, this verifies (1.23), and the proof is complete. $\square$

Similarly, it can be shown that

$$\lim_{n\to\infty} E \sup_{[0,T]} [X_n(t) - X_0(t)]^2 = 0$$

from which it follows that

$$\sup_{[0,T]} |X_n(t) - X_0(t)| \longrightarrow 0 \qquad \text{in probability.} \tag{1.27}$$

Furthermore, by employing the truncation procedure, one can obtain (1.27) when assumption 2 is replaced by the weaker condition

$$|X_n(0) - X_0(0)| \longrightarrow 0 \qquad \text{in probability.}$$

A number of other generalizations and variations of Theorem 3.5 are well known. The discrete parameter n can be replaced by a continuous parameter, say ν; the conclusion

$$\lim_{\nu\to\nu_0} \sup_{[0,T]} E[X_\nu(t) - X_{\nu_0}(t)]^2 = 0$$

indicates that the solution is mean square continuous with respect to the parameter ν. The initial condition $X_n(0)$ may be replaced by a random function $Z_n(t)$; in this case assumption 2 takes the form

$$\lim_{n\to\infty} \sup_{[0,T]} E[Z_n(t) - Z_0(t)]^2 = 0.$$

Higher-order dependence (e.g., differentiability) of solutions of stochastic equations with respect to parameters and initial conditions is possible, but more restrictive hypotheses are required to obtain the corresponding results. The following result is an example which can be easily stated.

THEOREM 3.6   Assume f and g satisfy the conditions of Theorem 3.1. For any real x, let $X(t; x)$ denote the solution of (0.9) with initial value $X(0) = x$, that is $X(t; x)$ satisfies (1.1) with $X_0 = x$,

$$X(t; x) = x + \int_0^t f(s, X(s; x)) \, ds + \int_0^t g(s, X(s; x)) \, dW(s).$$

Suppose further that the derivatives

$$\frac{\partial f(t, x)}{\partial x} \qquad \text{and} \qquad \frac{\partial g(t, x)}{\partial x}$$

are continuous and bounded. Then the derivative

$$Y(t) = \frac{\partial}{\partial x}[X(t; x)]$$

(in the mean square sense) exists and satisfies

$$Y(t) = 1 + \int_0^t \frac{\partial f}{\partial x}(s, X(s; x)) Y(s) \, ds$$

$$+ \int_0^t \frac{\partial g}{\partial x}(s, X(s; x)) Y(s) \, dW(s). \qquad (1.28)$$

The proof is indicated in Gihman and Skorohod (1972, p. 59).

Notice that (1.28) indicates that the process $Y(t)$ satisfies a stochastic differential equation where the coefficients have explicit dependence on the probability space variable ω as opposed to dependence on ω only via Y itself. This gives some indication of why it is important that the basic theory of stochastic equations, as exemplified by Theorems 3.1, 3.4, and 3.5, can be extended to the case of random functions $f(t, \omega, x)$ and $g(t, \omega, x)$. It is inappropriate here to consider ω simply as a parameter in the usual sense, since for fixed ω (1.1) is not meaningful as a deterministic equation, in general. That ω is not considered as a parameter distinguishes stochastic differential equations from other random equations. What is required to carry out the extension of the theory to the random case is some additional hypothesis on the explicit dependence of f and g on ω. It is clear that $f(t, \omega, x)$ and $g(t, \omega, x)$ must be nonanticipating random functions, at least. In addition to being measurable with respect to all variables, the nonanticipating condition requires that, for almost all fixed t and x, the random variables $f(t, \cdot, x)$ and $g(t, \cdot, x)$ are $\mathcal{A}(t)$-measurable. Conditions imposed on the deterministic functions f and g in the t and x variables, for example the Lipschitz and growth assumptions in the existence and uniqueness theorem, yield analogous results for the random function case if they are assumed to hold with probability 1. In some cases, requiring the conditions to be satisfied in probability suffices. For example, Theorem 3.5 is valid for random functions f_n and g_n if the measurability conditions indicated above are satisfied and assumption 1 is replaced by

$$\lim_{n \to \infty} P\big(\sup_{|x| \leq N} (|f_n(t, \omega, x) - f_0(t, \omega, x)|$$

$$+ |g_n(t, \omega, x) - g_0(t, \omega, x)|) > \varepsilon \big) = 0,$$

$$\text{for each } N > 0, t \in [0, T], \text{ and } \varepsilon > 0$$

[Gihman and Skorohod (1972), p. 52].

3.2 ITO'S FORMULA AND MOMENTS OF SOLUTIONS OF STOCHASTIC DIFFERENTIAL EQUATIONS

Ito's formula, the stochastic chain rule, is the most important tool of stochastic analysis as it gives the stochastic differential of a smooth function of the solution of a stochastic differential equation. This section parallels Section 2.4 of the previous chapter where Ito's formula was presented for functions of stochastic integrals, along with an application yielding an estimate for moments of stochastic integrals. In particular, Theorems 3.7 and 3.8 of this section are the counterparts, for solutions of stochastic differential equations, of Theorems 2.9 and 2.10 in Section 2.4.

THEOREM 3.7 Assume that the functions f and g satisfy the conditions of Theorem 3.1 guaranteeing the existence and uniqueness of solutions to initial value problems involving equation (0.9). If the real-valued function $F(t,x)$ has continuous partial derivatives $\partial F/\partial t$, $\partial F/\partial x$, and $\partial^2 F/\partial x^2$ for $t \in [0,T]$ and $x \in R$, and $X(t)$ is a solution of (0.9), then the process $F(t,X(t))$ has the stochastic differential

$$
dF(t,X(t)) = \left[\frac{\partial F}{\partial t} + f \frac{\partial F}{\partial x} + \frac{1}{2} g^2 \frac{\partial^2 F}{\partial x^2} \right] (t,X(t))\, dt
$$
$$
+ \left[\frac{\partial F}{\partial x} g \right] (t,X(t))\, dW(t). \tag{2.1}
$$

Proof: The result follows immediately from Theorem 2.9. □

EXAMPLE 3.2 Consider the linear stochastic differential equation

$$
dX(t) = f(t)X(t)\, dt + g(t)X(t)\, dW(t)
$$

where f and g are continuous functions, and let $F(x) = x^2$. Theorem 3.7 can be applied, and (2.1) becomes

$$
dX^2(t) = [2f(t)X^2(t) + g^2(t)X^2(t)]\, dt + 2g(t)X^2(t)\, dW(t);
$$

the processs $Y(t) = X^2(t)$ solves the linear equation

$$
dY = (2f(t) + g^2(t))Y\, dt + 2g(t)Y\, dW.
$$

The vector version of Theorem 3.7 is completely analogous: suppose $X(t)$ is a solution of the vector equation (0.10); writing $X = \{X_i\}$, $W = \{W_j\}$, $f = \{f_i\}$, and $G = \{g_{ij}\}$, $1 \le i \le n$, $1 \le j \le m$, one has, in place of

(2.1), written in component form,

$$dF(t, X(t))$$

$$= \left[\frac{\partial F}{\partial t} + \sum_{i=1}^{n} f_i \frac{\partial F}{\partial x_i} + \sum_{i,j=1}^{n} \sum_{k=1}^{m} \frac{1}{2} g_{ik} g_{jk} \frac{\partial^2 F}{\partial x_i \partial x_j} \right] (t, X(t)) \, dt$$

$$+ \sum_{i=1}^{n} \frac{\partial F}{\partial x_i} (t, X(t)) \sum_{j=1}^{m} g_{ij}(t, X(t)) \, dW_j(t), \qquad (2.2)$$

for scalar functions F with continuous partial derivatives $\partial F/\partial t$, $\partial F/\partial x_i$, and $\partial^2 F/\partial x_i \partial x_j$; in vector-matrix form (2.2) becomes

$$dF = [F_t + f^T F_x + \tfrac{1}{2} \operatorname{tr}(GG^T F_{xx})] \, dt + F_x^T G \, dW, \qquad (2.3)$$

where the t, x, and xx subscripts denote the scalar, vector, and matrix, respectively, partial derivatives (Here the superscript T denotes transpose, and tr denotes the trace on sum of main diagonal entries.) Ito's formula is valid for random functions f, G, and F, with hypotheses modified as indicated in the remarks at the end of the previous section. The main result of this section, which gives estimates for the even order moments of solutions similarly to the estimates for the moments of stochastic integrals given in Theorem 2.10 of the previous chapter, now follows.

THEOREM 3.8   Assume that the functions f and g satisfy the conditions of Theorem 3.1 guaranteeing the existence and uniqueness of solutions to initial value problems involving equation (0.9). If, for a positive integer n,

$$E|X_0|^{2n} < \infty,$$

then the solution $X(t)$ of (0.9) on $[0, T]$ with initial value $X(0) = X_0$ satisfies

$$E|X(t)|^{2n} \le (1 + E|X_0|^{2n}) e^{Ct} \qquad (2.4)$$

and

$$E|X(t) - X_0|^{2n} \le D(1 + E|X_0|^{2n}) t^n e^{Ct} \qquad (2.5)$$

for each $t \in [0, T]$ where $C = 2n(2n+1)K^2$ and D are constants (depending only on n, K, and T).

Proof: Let

$$f_N(t, x) = \begin{cases} f(t, x), & \text{if } |x| \le N \\ f(t, Nx/|x|), & \text{if } |x| > N \end{cases}$$

and define $g_N(t, x)$ similarly. Also, let

$$X_0^{(N)}(\omega) = \begin{cases} X_0(\omega), & \text{if } |X_0(\omega)| \le N \\ NX_0(\omega)/|X_0(\omega)|, & \text{if } |X_0(\omega)| > N. \end{cases}$$

The truncated process $X_N(t)$ which is the solution of the stochastic initial value problem

$$dX_N(t) = f_N(t, X_N(t))\, dt + g_N(t, X_N(t))\, dW(t) \tag{2.6}$$

$$X_N(0) = X_0^{(N)}$$

converges uniformly on $[0, T]$ w.p. 1 to $X(t)$ as $N \to \infty$. By applying Ito's formula to $F(t, x) = |x|^{2n}$ and $X_N(t)$, one obtains

$$|X_N(t)|^{2n} = |X_0^{(N)}|^{2n} + \int_0^t 2n |X_N(s)|^{2n-2} X_N(s) f_N(s, X_N(s))\, ds$$

$$+ \int_0^t n |X_N(s)|^{2n-2} |g_N(s, X_N(s))|^2\, ds$$

$$+ \int_0^t 2n(n-1) |X_N(s)|^{2n-4} |X_N(s) g_N(s, X_N(s))|^2\, ds$$

$$+ \int_0^t 2n |X_N(s)|^{2n-2} X_N(s) g_N(s, X_N(s))\, dW(s).$$

The expected value of the last integral is zero due to the boundedness of the integrand; so appealing to the growth condition,

$$E|X_N(t)|^{2n} = E|X_0^{(N)}|^{2n} + \int_0^t E([2X_N(s) f_N(s, X_N(s))$$

$$+ |g_N(s, X_N(s))|^2] n |X_N(s)|^{2n-2}$$

$$+ 2n(n-1) |X_N(s)|^{2n-4} |X_N(s) g_N(s, X_N(s))|^2)\, ds$$

$$\le E|X_0^{(N)}|^{2n} + (2n+1)nK^2$$

$$\times \int_0^t E([1 + |X_N(s)|^2] |X_N(s)|^{2n-2})\, ds. \tag{2.7}$$

Now, making use of the inequality $(1 + |x|^2)|x|^{2n-2} \le 1 + 2|x|^{2n}$, obtains from (2.7)

$$E|X_N(t)|^{2n} \le E|X_0^{(N)}|^{2n} + (2n+1)nK^2 t$$

$$+ (2n+1)2nK^2 \int_0^t E|X_N(s)|^{2n}\, ds.$$

By Lemma 3.2

$$E|X_N(t)|^{2n} \le h(t) + 2n(2n+1)K^2$$
$$\times \int_0^t \exp\left\{2n(2n+1)K^2(t-s)\right\} h(s)\, ds \tag{2.8}$$

where $h(t) = E|X_0^{(N)}|^{2n} + (2n+1)nK^2t$. Carrying out the integration in (2.8) yields (2.4) for $X_N(t)$. Letting $N \to \infty$ obtains the result for $X(t)$.

Instead of applying Ito's formula, it is convenient to work directly from the stochastic differential equation to establish (2.5).

$$E|X(t) - X_0|^{2n} = E\left|\int_0^t f(s, X(s))\, ds + \int_0^t g(s, X(s))\, dW(s)\right|^{2n}$$
$$\le 2^{2n-1}\left\{ E\left|\int_0^t f(s, X(s))\, ds\right|^{2n}\right.$$
$$\left. + E\left|\int_0^t g(s, X(s))\, dW(s)\right|^{2n}\right\}.$$

Applying Hölder's inequality and Theorem 2.10 results in the estimate

$$E|X(t) - X_0|^{2n} \le 2^{2n-1}[t^{2n-1}\int_0^t E|f(s, X(s))|^{2n}\, ds$$
$$+ t^{n-1}(n(2n-1))^n \int_0^t E|g(s, X(s))|^{2n}\, ds].$$

Thus, from the growth condition, one sees that there is a constant D_1 depending on t, n, and K such that

$$E|X(t) - X_0|^{2n} \le D_1 t^{n-1}\int_0^t E[1 + |X(s)|^2]^n\, ds$$
$$\le D_1 t^{n-1}\int_0^t [2^{n-1} + 2^{n-1}E|X(s)|^{2n}]\, ds$$
$$\le 2^{n-1}D_1 t^{n-1}\int_0^t [1 + (1 + E|X_0|^{2n})e^{Cs}]\, ds.$$
$$\le 2^n t^n D_1(1 + E|X_0|^{2n})e^{Ct}.$$

Specifically, $D_1 = 2^{2n-1}K^{2n}[t^n + (n(2n-1))^n]$, so one has

$$E|X(t) - X_0|^{2n} \le D(1 + E|X_0|^{2n})t^n e^{Ct},$$

where

$$D = 2^{3n-1} K^{2n} [t^n + (n(2n-1))^n]. \quad \square$$

3.3   SOLUTIONS OF STOCHASTIC DIFFERENTIAL EQUATIONS AS DIFFUSION PROCESSES

The mathematical properties of stochastic differential equations exhibited in Section 3.1 mainly reflect the differential equations character of these models. In this section the basic theory that stems primarily from the stochastic nature of the equations is presented. The main point is that solutions of stochastic differential equations represent arbitrary diffusion processes as transformations of the prototype diffusion process, namely, the Wiener process. The existence of a well-developed theory for diffusion processes, which to a large extent follows from the fact that the probability laws for such processes can be simply characterized, motivates establishing this result, and provides further evidence of the mathematical tractability of stochastic differential equations.

First of all, it is verified that solutions $X(t)$ of stochastic differential equations (0.9) are Markov processes. Recall from Chapter 1 (Section 1.7) that the probability law for such a process is completely determined by the initial distribution

$$P_0(B) = P(X(0) \in B) \tag{3.1}$$

and the transition probabilities

$$P(s, x, t, B) = P(X(t) \in B \mid X(s) = x) \tag{3.2}$$

where $x \in R$, $B \subseteq \mathcal{B}$, and $0 \leq s \leq t \leq T$. Note that, considering s and x fixed, the transition probability $P(s, x, t, B)$, given by (3.2), is precisely the distribution of the solution $X(t) = X(t; s, x)$ of the equation

$$X(t) = x + \int_s^t f(r, X(r)) \, dr + \int_s^t g(r, X(r)) \, dW(r). \tag{3.3}$$

THEOREM 3.9   Assume that the functions f and g satisfy the conditions of Theorem 3.1 guaranteeing the existence and uniqueness of solutions to initial value problems involving equation (0.9). Then, for arbitrary initial values $X_0 = X(0)$, the solution $X(t)$ of (0.9) is a Markov process on the interval $[0, T]$ with initial distribution and transition probabilities given by (3.1) and (3.2) respectively.

Proof: As usual, denote by $(\Omega, \mathcal{A}, P)$ the underlying probability space; let $\mathcal{W}(t)$ represent the sub-σ-algebras of $\mathcal{A}$ generated by X_0 and $W(s)$, $0 \le s \le t$, and let

$$\mathcal{A}(t) = \mathcal{A}[0,t] = \mathcal{A}(X(s), 0 \le s \le t) \tag{3.4}$$

the σ-algebra generated by the solution $X(t)$ up to time t. To verify the Markov property here one needs to show (7.2) of Section 1.7, namely, for $0 \le s \le t \le T$ and $B \in \mathcal{B}$,

$$P(X(t) \in B \mid \mathcal{A}(s)) = P(X(t) \in B \mid X(s)) \text{ w.p. 1.} \tag{3.5}$$

In what follows, properties of conditional expectation given in Section 1.4 of Chapter 1 are invoked.

To begin, since $X(t)$ is $\mathcal{W}(t)$-measurable,

$$\mathcal{A}(t) \subseteq \mathcal{W}(t),$$

so (3.5) is implied by establishing the stronger condition

$$P(X(t) \in B \mid \mathcal{W}(s)) = P(X(t) \in B \mid X(s)). \tag{3.6}$$

To verify (3.6), it suffices to prove: for every bounded measurable function $h(x, \omega)$ defined on $R \times \Omega$ of the form

$$h(x, \omega) = \sum_{i=1}^{n} h_i(x) H_i(\omega) \tag{3.7}$$

where each H_i is a random variable independent of $\mathcal{W}(s)$,

$$E(h(X(s), \omega) \mid \mathcal{W}(s)) = E(h(X(s), \omega) \mid X(s)). \tag{3.8}$$

This would imply that (3.8) holds for the class of all bounded measurable functions $h(x, \omega)$ for which the random variables $h(x, \cdot)$ are independent of $\mathcal{W}(s)$, since the subclass defined by (3.7) is dense in this class. In particular, then, (3.8) would hold for the choice

$$h(x, \omega) = I_B(X(t; s, x)) \tag{3.9}$$

where $I_B(X)$ denotes the indicator function that $X \in B$, $B \in \mathcal{B}$; this follows since I_B is obviously bounded and inherits measurability and independence of $\mathcal{W}(s)$ from $X(t; s, x)$ which is measurable with respect to the σ-algebra generated by the increment $W(t) - W(s)$. Together with the basic semigroup property

$$X(t; 0, X_0) = X(t; s, X(s; 0, X_0)), \qquad \text{all } 0 \le s \le t \le T,$$

the choice (3.9) in (3.8) coincides with (3.6) and so the theorem is proved once (3.8) is verified for functions of the form (3.7).

But this follows immediately from the properties of conditional expectation, since $X(s)$ is certainly $\mathcal{A}(s)$-measurable and each H_i is independent of $X(s)$ and thus

$$E(h(X(s),\omega)) \mid \mathcal{W}(s)) = \sum_{i=1}^{n} h_i(X(s))E(H_i) = E(h(X(s),\omega) \mid X(s)).$$

and so the proof is complete. $\square$

Recall that a Markov process is said to be homogeneous if the transition probabilities are stationary,

$$P(s+u,x,t+u,B) = P(s,x,t,B), \qquad \text{all } 0 \le u \le T-t;$$

that is, for fixed x and b the transition probability is a function of $t-s$ only

$$P(s,x,t,B) = P(t-s,x,B),$$

which affects a further simplification of the probability law description. It is easy to see that if the stochastic differential equation is autonomous, that is, $f(t,x) = f(x)$ and $g(t,x) = g(x)$, then the solution will be a homogenous Markov process.

Diffusion processes are Markov processes whose probability law is specified by the drift and diffusion coefficients, two functions which correspond to the conditional infinitesimal mean and variance of the process respectively. This is pointed out in Section 1.8 of Chapter 1, in noting that the transition probability density is a solution of the backward and forward partial differential equations determined by the drift and diffusion coefficients. For stochastic differential equations (0.9), in particular, one has the following theorem.

THEOREM 3.10   Assume that the functions f and g satisfy the conditions of Theorem 3.1 guaranteeing the existence and uniqueness of solutions to initial value problems involving equation (0.9). Then any solution $X(t)$ of (0.9) is a diffusion process on the interval $[0,T]$ with drift coefficient $f(t,x)$ and diffusion coefficient $g^2(t,x)$.

Proof: Properties (8.1) to (8.3) in Chapter 1, Section 1.8, need to be verified. First, if it can be shown that, for some $\delta > 0$,

$$\lim_{t \downarrow s} \frac{1}{t-s} \int_R |y-x|^{2+\delta} P(s,x,t,dy) = 0; \tag{3.10}$$

then (8.1) of Chapter 1, Section 1.8, can be verified, as

$$\int_{[y-x]>\varepsilon} P(s,x,t,dy) \le \varepsilon^{-2-\delta} \int_R |y-x|^{2+\delta} P(s,x,t,dy).$$

But that (3.10) holds for $\delta = 2$ follows from

$$E_{s,x}|X(t)-X(s)|^4 \le C(t-s)^2 \qquad (3.11)$$

for some constant C and $0 \le s \le t \le T$; (3.11) can be seen from Theorem 3.8.

It remains to verify properties (8.2) and (8.3) which can be written; As $t \downarrow s$,

$$E(X(t;s,x)-x) = f(s,x)(t-s) + o(t-s) \qquad (3.12)$$

and

$$E|X(t;s,x)-x|^2 = g^2(s,x)(t-s) + o(t-s). \qquad (3.13)$$

These are verified similarly; here the proof of (3.12) is given. Using the conditions for existence and uniqueness of solutions one has

$$E(X(t;s,x)-x) = \int_s^t Ef(u,X(u;s,x))\,du$$
$$(3.14)$$
$$= \int_s^t f(u,x)\,du + \int_s^t E(f(u,X(u;s,x))-f(u,x))\,du.$$

Invoking the Cauchy-Schwartz inequality and the Lipschitz condition, one obtains

$$\left| \int_s^t E(f(u,X(u;,s,x))-f(u,x))\,du \right|$$

$$\le \int_s^t E|f(u,X(u;s,x))-f(u,x)|\,du$$

$$(3.15)$$

$$\le (t-s)^{1/2} \left(\int_s^t E|f(u,X(u;s,x))-f(u,x)|^2\,du \right)^{1/2}$$

$$\le O((t-s)^{1/2}) \left(\int_s^t E|X(u;s,x)-x|^2\,du \right)^{1/2}.$$

From Theorem 3.8, the last factor in (3.15) is $O(t-s)$, and so, as $t \downarrow s$,

$$\left| \int_s^t E(f(u,X(u;s,x))-f(u,x))\,du \right| \le (t-s)^{3/2}O(1). \qquad (3.16)$$

Continuity of f in the first variable allows

$$\int_s^t f(u,x)\,du = f(s,x)(t-s) + \int_s^t f(u,x) - f(t,x)\,du \tag{3.17}$$
$$= f(s,x)(t-s) + o(t-s).$$

Then (3.14), (3.16), and (3.17) together imply (3.12). Equation (3.13) follows similarly, and it is left as an exercise (Exercise 3.7). □

In the vector case, the drift coefficient of the process X is a vector function $f(t,x)$ and the diffusion coefficient is a matrix function $G(t,x)G(t,x)^T$. Now attention is turned to the converse question: given a diffusion process $Y(t)$ defined on some interval, under what conditions will $Y(t)$ satisfy a stochastic differential equation? A weak form of this question is easily answered. Suppose $a(t,y)$ and $b(t,y)$ are the drift and diffusion coefficients of the given process $Y(t)$, and a and $\sqrt{b}$ satisfy the existence and uniqueness conditions in Theorem 3.1. If $W = W(t)$ is any standard Wiener process, the solution $X(t)$ of the initial value problem

$$dX = a(t,X)dt + \sqrt{b(t,X)}\,dW$$
$$X(t_0) = Y(t_0) \tag{3.18}$$

shares the same probability law, and hence is equivalent to the given process $Y(t)$. However, the sample paths of X and Y may not coincide with probability 1. (Note that the Wiener process was chosen arbitrarily as well as the positive square root of b; in the vector case, the dimension of W which coincides with the column dimension of the matrix G can be chosen arbitrarily in addition to the specific choice of G so that GG^T is the given diffusion coefficient matrix.) To deal with the stronger version of this question, the following scheme can be employed which obtains a stochastic differential equation having the given diffusion $Y(t)$ as a solution:

Let $Z(t) = g(t,Y(t))$ where

$$g(t,y) = \int_0^y \frac{dv}{\sqrt{b(t,v)}}; \tag{3.19}$$

define the function $\bar{a}(t,z)$ by

$$\bar{a}(t,z) = \left[\frac{\partial g}{\partial t} + a\frac{\partial g}{\partial y} + \frac{1}{2}b\frac{\partial^2 g}{\partial y^2}\right](t,g^{-1}(t,z)); \tag{3.20}$$

then Z is a diffusion process with drift $\bar{a}(t,z)$ and diffusion coefficient unity. If the process W_y is defined by

$$W_y(t) = Z(t) - Z(0) - \int_0^t \bar{a}(s, Z(s))\, ds, \tag{3.21}$$

one shows that W_y is a Wiener process. Notice that if W_y is a Wiener process, (3.21) is equivalent to the stochastic differential equation

$$dZ(t) = \bar{a}(t, Z(t))\, dt + dW_y(t),$$

and by (3.19), (3.20), using Ito's formula, this implies that $Y(t)$ is a solution of the stochastic equation

$$dY(t) = a(t, Y(t))\, dt + \sqrt{b(t, Y(t))}\, dW_y(t). \tag{3.22}$$

The next theorem gives conditions on the coefficients $a(t,y)$ and $b(t,y)$ of the given diffusion process $Y(t)$ which enable verification that the process $W_y(t)$, constructed in (3.21), is a Wiener process. First, though, the assertion that $Z(t)$ is a diffusion process with the indicated drift and diffusion coefficients is established via a change of variables argument.

Since g is monotone in y, the process $Z(t)$ defined by (3.19) is a Markov process; its transition function $\tilde{P}(s, z, t, A)$ is given in terms of the transition function $P(s, y, t, B)$ of the diffusion process $Y(t)$ by

$$\tilde{P}(s, z, t, A) = P(s, g^{-1}(s, z), t, g^{-1}(t, A))$$

where $g^{-1}(s, z)$ is the inverse of $g(s, y)$ with respect to y and

$$g^{-1}(t, A) = \{y\colon g(t, y) \in A\}\,.$$

Then with $y = g^{-1}(s, z)$ and $\eta = g^{-1}(t, \varsigma)$, one has

$$\underset{t\downarrow s}{\text{limit}}\, \frac{1}{t-s} \int\limits_{|z-\varsigma|>\varepsilon} P(s, g^{-1}(s, z), t, g^{-1}(t, d\varsigma))$$

$$= \underset{t\downarrow s}{\text{limit}}\, \frac{1}{t-s} \int\limits_{|g(s,y)-g(t,\eta)|>\varepsilon} P(s, y, t, d\eta)$$

$$= 0.$$

(Here the continuous invertibility of g in the second variable allows replacing the set

$$\{\eta\colon |g(s, y) - g(t, \eta)| > \varepsilon\} \qquad \text{by} \qquad \{\eta\colon |y - \eta| > \delta\}$$

for some $\delta > 0$ provided t is sufficiently close to s.) The process $Z(t)$ satisfies condition (8.1) of Chapter 1 required of a Markov process to be a diffusion. Now suppose $g(t,y)$ is continuously differentiable in t and twice continuously differentiable in y. [These conditions are met for g given by (3.19), if assumption 2 concerning the function b in the next theorem holds.] Now it follows that $Z(t)$ inherits from $Y(t)$ the other two properties characterizing diffusions, namely, (8.2) and (8.3) in Chapter 1. To see this, observe that by Taylor's theorem, there are numbers θ, $\theta_1 \in (0,1)$ such that

$$\frac{1}{t-s} \int\limits_{|z-\varsigma|\leq\varepsilon} (\varsigma - z)P(s,g^{-1}(s,z),t,g^{-1}(t,d\varsigma))$$

$$= \frac{1}{t-s} \int\limits_{|g(s,y)-g(t,\eta)|\leq\varepsilon} [g(t,\eta) - g(s,y)]P(s,y,t,d\eta)$$

$$= \frac{\partial g(s + \theta(t-s),y)}{\partial s} \int\limits_{|g(s,y)-g(t,\eta)|\leq\varepsilon} P(s,y,t,d\eta) \qquad (3.23)$$

$$+ \frac{1}{t-s} \int\limits_{|g(s,y)-g(t,\eta)|\leq\varepsilon} \frac{\partial g(s,y)}{\partial y}(\eta - y)P(s,y,t,d\eta)$$

$$+ \frac{1}{t-s} \int\limits_{|g(s,y)-g(t,\eta)|\leq\varepsilon} \frac{1}{2}\frac{\partial^2 g(s,y + \theta_1(\eta - y))}{\partial^2 y}(\eta - y)^2 P(s,y,t,d\eta)$$

For the first integral on the right in (3.23), one has that

$$\int\limits_{|g(s,y)-g(t,\eta)|\leq\varepsilon} P(s,y,t,d\eta) = 1 - \int\limits_{|g(s,y)-g(t,\eta)|>\varepsilon} P(s,y,t,d\eta)$$

$$= 1 + o(t-s), \qquad \text{as } t \longrightarrow s.$$

By choosing ε sufficiently small, the factor

$$\frac{\partial^2 g(s,y + \theta_1(\eta - y))}{\partial y^2}$$

can be made arbitrarily close to $\partial^2 g(s,y)/\partial y^2$. So noting that the drift and diffusion coefficients of $Y(t)$ are $a(t,y)$ and $\sqrt{b(t,y)}$ respectively and taking the limit of the expression on the right in (3.23) as $t \downarrow s$ yields

$$\left[\frac{\partial g}{\partial t} + \frac{\partial g}{\partial y}a + \frac{1}{2}\frac{\partial^2 g}{\partial y^2}b\right](s,g^{-1}(s,z)). \qquad (3.24a)$$

Similarly, one obtains

$$\underset{t\downarrow s}{\text{limit}}\ \frac{1}{t-s}\int\limits_{|z-\delta|\le\varepsilon}(\varsigma-z)^2 P(s,g^{-1}(s,z),t,g^{-1}(t,d\varsigma))$$

$$=\left[\left(\frac{\partial g}{\partial x}\right)^2 b\right](s,g^{-1}(s,z)).$$

(3.24b)

For the particular choice of g given by (3.19) the functions (3.24a) and (3.24b) are respectively

$$\bar{a}(s,z)=-\frac{1}{2}\int_0^{g^{-1}(s,z)}\left[b^{-(3/2)}\frac{\partial b}{\partial t}\right](s,v)\,dv$$

$$+\frac{1}{4\sqrt{b}}\left[4a-\frac{\partial b}{\partial y}\right](s,g^{-1}(s,z)),$$

and   (3.25)

$$\bar{b}(s,z)\equiv 1$$

where $g^{-1}(s,z)$ is given implicitly by the equation

$$z=\int_0^{g^{-1}(s,z)}\frac{1}{\sqrt{b(t,v)}}\,dv.$$

Thus $Z(t)$ is a diffusion process with coefficients given by (3.25).

THEOREM 3.11   Let $Y(t)$ be a diffusion process on $[0,T]$ with coefficients $a(t,y)$ and $b(t,y)$ satisfying the following conditions for all y and $0\le s\le t\le T$:

1. $a(t,y)$ is continuous in both variables and for some constant K, $|a(t,y)|\le K(1+|y|)$.
2. $b(t,y)$ is continuous in both variables, $1/b$ is bounded, and the partial derivatives $\partial b/\partial t$ and $\partial b/\partial x$ are continuous and bounded.
3. There is a function $\psi(y)>1+|y|$ such that
   a. $\sup_{[0,T]}E\psi(Y(t))<\infty$.
   b. $E_{s,y}|Y(t)-Y(s)|+E_{s,y}|Y(t)-Y(s)|^2\le\psi(y)(t-s)$.
   c. $E_{s,y}|Y(t)|+E_{s,y}|Y(t)|^2\le\psi(y)$.

Then there is a Wiener process $W_y(t)$ for which $Y(t)$ solves the stochastic differential equation

$$dY(t)=a(t,Y(t))\,dt+\sqrt{b(t,Y(t))}\,dW_y(t).$$

(3.26)

Proof: To show that W_y given by (3.21) is a Wiener process, the following lemma is employed.

LEMMA 3.12   Let the process $B(t)$ be defined on $[0,T]$ and let $\mathcal{B}(t)$ denote the σ-algebra generated by the variables $B(s)$, $s \leq t$. Suppose further that the following conditions are satisfied:

1. $B(t)$ is continuous w.p. 1 and $B(0) = 0$.
2. There exist finite mean random variables M_1 and M_2 with
    (a) $(1/h)E(B(t+h) - B(t) \mid \mathcal{B}(t)) \leq M_1$.
    (b) $(1/h)E(|B(t+h) - B(t)|^2 \mid \mathcal{B}(t)) \leq M_2$.
3. For each $t \in [0,T]$, w.p. 1,
    (a) $\text{limit}_{h \to 0}(1/h)E(B(t+h) - B(t) \mid \mathcal{B}(t)) = 0$.
    (b) $\text{limit}_{h \to 0}(1/h)E(|B(t+h) - B(t)|^2 \mid \mathcal{B}(t)) = 1$.

Then $B(t)$ is a (standard) Wiener process.

Proof of the lemma: For $t > s$, define

$$\varphi(t) = E(B(t) - B(s) \mid \mathcal{B}(s)).$$

Then, one has, from the properties of conditional expectation,

$$\underset{h \downarrow 0}{\text{limit}}\, \frac{\varphi(t+h) - \varphi(t)}{h} = \underset{h \downarrow 0}{\text{limit}}\, E\left(\frac{B(t+h) - B(t)}{h} \,\middle|\, \mathcal{B}(s) \right) \qquad (3.27)$$

$$= \underset{h \downarrow 0}{\text{limit}}\, E\left\{ E\left(\frac{B(t+h) - B(t)}{h} \,\middle|\, \mathcal{B}(t) \right) \,\middle|\, \mathcal{B}(s) \right\}.$$

Furthermore, the dominated convergence theorem for conditional expectation can be applied since 2a holds; thus the limit and conditional expectation can be interchanged in (3.27), which means that

$$\underset{h \downarrow 0}{\text{limit}}\, \frac{\varphi(t+h) - \varphi(t)}{h} = 0. \qquad (3.28)$$

It is clear that $\varphi(t)$ is continuous w.p. 1. Together with (3.28), this implies that φ is constant. Since

$$\underset{t \downarrow s}{\text{limit}}\, \varphi(t) = 0,$$

then $\varphi(t) \equiv 0$; that is, for $t \geq s$,

$$E(B(t) - B(s) \mid \mathcal{B}(s)) = 0. \qquad (3.29)$$

Similarly, defining

$$\bar{\varphi}(t) = E(|B(t) - B(s)|^2 \mid \mathcal{B}(s))$$

one obtains

$$\underset{h\downarrow 0}{\text{limit}}\ \frac{\bar{\varphi}(t+h)-\bar{\varphi}(t)}{h}=1,$$

from which it follows that $\bar{\varphi}(t)=t-s,\ t\geq s$; that is,

$$E(|B(t)-B(s)|^{2}\mid\mathcal{B}(s))=t-s. \qquad (3.30)$$

The lemma now follows by Doob's theorem (Theorem 1.8, Section 1.9). $\square$

Returning to the proof of the theorem, it suffices to show that $W_{y}(t)$ satisfies the conditions of the lemma. It is clear that $W_{y}(t)$ satisfies condition 1. Toward verifying 2a consider the process $Z(t)$ defined by (3.19).
By Taylor's formula, one has

$$E(Z(t+h)-Z(t)\mid Z(t))=E(g(t+h,Y(t+h))-g(t,Y(t))\mid Y(t))$$

$$=E\left(\frac{\partial g}{\partial t}(t+\theta h,Y(t))\mid Y(t)\right)h$$

$$+\frac{\partial g}{\partial y}(t,Y(t))E(Y(t+h)-Y(t)\mid Y(t))$$

$$+\frac{1}{2}E\left(\frac{\partial^{2}g}{\partial y^{2}}(t,Y(t))+\theta'[Y(t+h)-Y(t)]\right)$$

$$\times|Y(t+h)-Y(t)|^{2}\mid Y(t)\right), \qquad (3.31)$$

where θ and θ' are in $(0,1)$. By the assumption 2 on the function b, and the definition of g in (3.19), one has that $|g(t,y)|\leq C|y|$, for some constant C, and that $\partial g/\partial y$ and $\partial^{2}g/\partial y^{2}$ are bounded. These considerations, together with assumption 3, and (3.31) yield the estimate, for some constant C_{1},

$$|E(Z(t+h)-Z(t)\mid Z(t))|\leq C_{1}\bar{\psi}(t,Z(t))h \qquad (3.32)$$

where

$$\bar{\psi}(t,z)=\psi(g^{-1}(t,z)).$$

By 3a

$$\underset{[0,T]}{\text{sup}}\ E\bar{\psi}(t,Z(t))=\underset{[0,T]}{\text{sup}}\ E\psi(Y(t))<\infty.$$

Let $\mathcal{A}(t)$ denote the σ-algebra generated by $Z(s)$, $s \leq t$. The from (3.21), $W_y(t)$ is measurable with respect to $\mathcal{A}(t)$ and

$$|E(W_y(t+h) - W_y(t) \mid \mathcal{A}(t))|$$

$$= |E(Z(t+h) - Z(t)|Z(t)) - \int_t^{t+h} E(\bar{a}(s, Z(s))|\mathcal{A}(t)) \, ds|$$

$$\leq C_1 \bar{\psi}(t, Z(t))h + C_2 \int_t^{t+h} E(1 + |Y(s)| \mid \mathcal{A}(t)) \, ds \qquad (3.33)$$

$$\leq C_3(\bar{\psi}(t, Z(t)) + 1)h$$

where C_2 and C_3 are constants; in obtaining (3.33), (3.32), (3.19), (3.20), and all three asssumptions of the theorem are used.

Since

$$\lim_{h \to 0} \int_t^{t+h} E(\bar{a}(s, Z(s)) \mid \mathcal{A}(t)) \, ds = \bar{a}(t, Z(t)),$$

and $\bar{a}(t, z)$ is the drift coefficient of the diffusion process $Z(t)$, it follows from (3.21) that

$$\lim_{h \to 0} E(W_y(t+h) - W_y(t) \mid \mathcal{A}(t)) = 0. \qquad (3.34)$$

Similarly to (3.33) and (3.34), one establishes

$$E(|W_y(t+h) - W_y(t)|^2 \mid \mathcal{A}(t)) \leq C_0 \bar{\psi}(t, Z(t))h \qquad (3.35)$$

and

$$\lim_{h \to 0} \frac{1}{h} E(|W_y(t+h) - W_y(t)|^2 \mid \mathcal{A}(t)) = 1. \qquad (3.36)$$

Hypotheses 2 and 3 of the lemma are verified for $B(t) = W_y(t)$ by (3.33) to (3.36). Thus W_y is a Wiener process and the proof is complete. $\square$

3.4 STRATONOVICH'S STOCHASTIC INTEGRAL

The ambiguity of the stochastic differential equation

$$dX(t) = f(t, X(t)) \, dt + G(t, X(t)) \, dW(t) \qquad (4.1)$$

disappears when the specific interpretation of the stochastic integral

$$\int G(u, X(u)) \, dW(u) \qquad (4.2)$$

involved in the integral counterpart of (4.1) is declared. As indicated in Section 2.1 there are infinitely many possible choices for the interpretation of (4.2). However, only two, the stochastic integrals of Ito and Stratonovich, have gained wide acceptance in the theory and applications literature. In this section the definition and basic properties of the Stratonovich stochastic integral are given. It is pointed out, in particular, that distinct diffusion processes are defined by the Ito and Stratonovich versions of (4.1). The theorem presented here is a slight extension of Stratonovich's result (1966) and also generalizes Theorem 2.3. For clarity, first, the one-dimensional situation is discussed: a one-parameter family of stochastic integrals, containing the Stratonovich as well as the Ito integral, is defined and a basic theorem is given which relates each integral in the family to the Ito integral.

Suppose $X(t)$ is a solution of the scalar stochastic differential equation

$$dX(t) = f(t, X(t))\, dt + g(t, X(t))\, dW(t) \tag{4.3}$$

defined on the time interval $[a, b]$, where the Ito interpretation of the term

$$g(t, X(t))\, dW(t)$$

is taken; that is, $X(t)$ describes a real diffusion process on $[a, b]$ with continuous drift f and diffusion g^2 coefficients. Let $h(t, x)$ be a real-valued function defined for all t in $[a, b]$, $x \in R$, which is continuous in t, and whose partial $\partial h / \partial x$ is continuous in t and x. Assume further that the process $h(t, X(t))$ is in the space $\mathcal{L}^2$ defined in Section 2.2; in particular, suppose that

$$\int_a^b E[h^2(t, X(t))]\, dt < \infty. \tag{4.4}$$

Consider partitions $a = t_0 < t_1 < \cdots < t_N = b$ of the interval $[a, b]$ with mesh size $\delta = \max(t_{i+1} - t_i)$. (Such a partition will be referred to as a δ-partition.) The family of stochastic integrals, indexed by the parameter λ, is defined by

$$\int_a^b h(t, X(t))\, dW(t) = \lim_{\delta \to 0} \sum_{i=0}^{N-1} h(t_i, \lambda X(t_{i+1}) + (1 - \lambda) X(t_i))\, \Delta W_i,$$
$$(\lambda)$$
$$\Delta W_i = W(t_{i+1}) - W(t_i), \tag{4.5}$$
$$\lambda \in [0, 1]$$

where the limit in (4.5) is in the mean square sense. (See Chapter 2.) Note, as in the example in Section 2.1, the $\lambda = 0$ case of (4.5) corresponds to the Ito integral; if $\lambda = \frac{1}{2}$, (4.5) defines the Stratonovich integral. In this section, for emphasis and clarity, the notation

$$\oint_a^b h(t, X(t))\, dW(t)$$

will be employed. Observe, from (4.5), that if $h(t)$ is a random function in the class $\mathcal{L}^2$, independent of X, then all integrals in this family, in particular the Ito and Stratonovich integrals, are identical. The example discussed in Section 2.1 corresponds to the special case of (4.5) where $X(t) = W(t)$ and $h(t,x) = x$; the right-hand side of (4.5) is shown to be

$$\tfrac{1}{2}[W^2(b) - W^2(a)] + (\lambda - \tfrac{1}{2})(b - a) \tag{4.6}$$

here, and in particular the Stratonovich case yields

$$\oint_a^b W(t)\, dW(t) = \tfrac{1}{2}[W^2(b) - W^2(a)]. \tag{4.7}$$

Equation (4.7) suggests that the Stratonovich integral may satisfy the usual rules of calculus. This point is substantiated after the next theorem.

THEOREM 3.13   Under the conditions stated above on the function h and the process X, for each $\lambda \in [0,1]$, the stochastic integral

$$\int_{\substack{a \\ (\lambda)}}^b h(t, X(t))\, dW(t)$$

exists, and satisfies

$$\int_{\substack{a \\ (\lambda)}}^b h\, dW = \int_a^b h\, dW + \lambda \int_a^b \frac{\partial h}{\partial x} g\, dt, \tag{4.8}$$

where the integrals on the right in (4.8) are the corresponding Ito integral and the ordinary integral along sample paths, respectively.

Proof: Note first that (from Section 2.2) under the conditions on h and X, the Ito integral

$$\int_a^b h(t, X(t))\, dW(t)$$

exists, and is the mean square limit of the sums

$$\sum_{i=0}^{N-1} h(t_i, X(t_i))[W(t_{i+1}) - W(t_i)] \tag{4.9}$$

as the mesh size $\delta \to 0$. Therefore, to establish (4.8) it suffices to show that some w.p. 1 limit of the difference of the sums in (4.5) and (4.9) agrees with the ordinary sample path integral in (4.8); that is, for some sequence of partitions with $\delta \to 0$,

$$\sum_{i=0}^{N-1} [h(t_i, \lambda X(t_{i+1}) + (1-\lambda)X(t_i)) - h(t_i, X(t_i))] \Delta W_i$$

$$\longrightarrow \lambda \int_a^b \left[\frac{\partial h}{\partial x} g\right](t, X(t))\, dt \qquad \text{w.p. 1}$$

(4.10)

where, once again, the notation $\Delta W_i = W(t_{i+1}) - W(t_i)$ is employed. Toward verifying (4.10), one has, for some θ_i, $0 \leq \theta_i \leq \lambda$,

$$h(t_i, \lambda X(t_{i+1}) + (1-\lambda)X(t_i)) - h(t_i, X(t_i)) = \lambda \frac{\partial h}{\partial x}\bigg|_{\theta_i} \Delta X_i \qquad (4.11)$$

where

$$\frac{\partial h}{\partial x}\bigg|_{\theta_i} = \frac{\partial h}{\partial x}(t_i, (1-\theta_i)X(t_i) + \theta_i X(t_{i+1})).$$

Consequently, to prove the result it suffices to show

$$\sum_{i=0}^{N-1} \frac{\partial h}{\partial x}\bigg|_{\theta_i} \Delta X_i \, \Delta W_i \longrightarrow \int_a^b \left[\frac{\partial h}{\partial x} g\right](t, X(t))\, dt \qquad \text{w.p. 1.} \qquad (4.12)$$

for some sequence of partitions of the interval $[a, b]$ as the mesh size $\delta \to 0$.

The following result of Goldstein (1969), which can be viewed as a generalization of Lemma 2.1, is useful to establish (4.12).

Suppose $Y(t)$ is a solution of the (in general, multidimensional) Ito equation

$$dY(t) = \tilde{f}(t, Y(t))\, dt + \tilde{G}(t, Y(t))\, d\tilde{W}(t) \qquad (4.13)$$

on the interval $[a, b]$: the vector- and matrix-valued functions $\tilde{f}$ and $\tilde{G}$, respectively, are assumed to satisfy the usual conditions guaranteeing existence and uniqueness of solutions of initial value problems involving (4.13). Then

$$\sum_{i=0}^{N-1} (\Delta Y_i)(\Delta Y_i)^T \longrightarrow \int_a^b [\tilde{G}\tilde{G}^T](t, Y(t))\, dt \qquad (4.14)$$

in probability as the mesh size δ of the corresponding partitions of the interval $[a, b]$ tends to zero.

Thus it follows that there is some sequence of partitions with $\delta \to 0$ such that the matrix limit statement (4.14) holds w.p. 1.

To apply (4.14) here, take $\tilde{W}(t) = W(t)$, and

$$Y(t) = \begin{bmatrix} X(t) \\ W(t) \end{bmatrix}.$$

Then, with $y = \left(\begin{smallmatrix} x \\ w \end{smallmatrix}\right)$, one has

$$\tilde{f}(t,y) = \begin{bmatrix} f(t,x) \\ 0 \end{bmatrix} \qquad \text{and} \qquad \tilde{G}(t,y) = \begin{bmatrix} g(t,x) \\ 1 \end{bmatrix},$$

so that

$$[\tilde{G}\tilde{G}^T](t,y) = \begin{bmatrix} g^2(t,x) & g(t,x) \\ g(t,x) & 1 \end{bmatrix}.$$

The off-diagonal components of (4.14), in this case, take the form

$$\sum_{i=0}^{N-1} \Delta X_i\, \Delta W_i \longrightarrow \int_a^b g(t, X(t))\, dt. \tag{4.15}$$

Then (4.12) follows from (4.15) by taking an appropriate sequence of partitions and applying the bounded convergence theorem along sample paths. $\square$

Suppose now that the function g satisfies the conditions mentioned above on h. One may consider the Stratonovich stochastic differential equation

$$d_S X(t) = f(t, X(t))\, dt + g(t, X(t))\, dW(t); \tag{4.16}$$

a solution $X(t)$ of (4.16) on an interval $[a, b]$ satisfies

$$\begin{aligned} X(t) - X(a) &= \int_a^t f(s, X(s))\, ds + \oint_a^t g(s, X(s))\, dW(s) \\ &= \int_a^t [f(s, X(s)) + \frac{1}{2}\left[\frac{\partial g}{\partial x} g\right](s, X(s))]\, ds \\ &\quad + \int_a^t g(s, X(s))\, dW(s) \end{aligned} \tag{4.17}$$

[by (4.8)] where the last integral is the Ito integral. Thus $X(t)$ is also a solution of the Ito stochastic differential equation

$$dX(t) = \left[f + \frac{1}{2}\frac{\partial g}{\partial x} g\right](t, X(t))\, dt + g(t, X(t))\, dW(t). \tag{4.18}$$

Similarly, a solution $X(t)$ of the Ito equation

$$dX = f(t, X)\, dt + g(t, X)\, dW \tag{4.19}$$

solves the Stratonovich equation

$$d_S X = \left[f - \frac{1}{2} \frac{\partial g}{\partial x} g \right] (t, X)\, dt + g(t, X)\, dW. \tag{4.20}$$

The next example extends (4.7) and so gives further indication that the calculus associated with the Stratonovich integral coincides with ordinary calculus. Specifically, it is shown that the usual chain rule for differentials

$$dh = h'(x)\, dx \tag{4.21}$$

holds for Stratonovich stochastic differentials in a special case.

EXAMPLE 3.3   Suppose the functions $g(x)$ and $h(x)$ are continuously differentiable and twice continuously differentiable respectively. Let $X(t)$ be a solution of the Stratonovich equation

$$d_S X(t) = g(X(t))\, dW(t). \tag{4.22}$$

Then it follows that

$$d_S h(X(t)) = h'(X(t))\, d_S X(t). \tag{4.23}$$

Toward verifying (4.23), first of all, note that $X(t)$ must solve the equivalent Ito equation

$$dX(t) = \tfrac{1}{2}[g'g](X(t))\, dt + g(X(t))\, dW(t), \tag{4.24}$$

from (4.18). Ito's formula, then, gives the stochastic differential for $h(X(t))$

$$dh(X(t)) = \tfrac{1}{2}[h'g'g + h''g^2](X(t))\, dt + [h'g](X(t))\, dW(t). \tag{4.25}$$

Making use of (4.8), the Ito term

$$[h'g](X(t))\, dW(t)$$

in (4.25) can be replaced by the expression

$$[h'g](X(t))\, dW(t) - \tfrac{1}{2}[(h'g)'g](X(t))\, dt, \tag{4.26}$$

with the Stratonovich interpretation. Substituting (4.26) into (4.25) and simplifying leads to

$$d_S h(X(t)) = [h'g](X(t))\, dW(t)$$

which, noting (4.22), confirms (4.23).

It is not difficult to show that (4.23) holds for the general scalar case, that is, for $h(t, x)$ with $\partial h/\partial t$ and $\partial^2 h/\partial x^2$ continuous, and $X(t)$, a solution of (4.16) with $\partial g/\partial x$ continuous (Exercise 3.9).

Distinct diffusion processes are described by the different interpretations of (4.3). For each λ the solution processes have the same diffusion coefficient, namely, g^2, but the diffusion process defined by the Stratonovich version has drift $f + (1/2)(\partial g/\partial x)g$, while the Ito equation defined process has drift f, and in general the drift coefficient is $f + \lambda(\partial g/\partial x)g$. In this sense the coefficients of the Ito equation coincide more nicely with the infinitesimal statistics of the diffusion process than those of the Stratonovich equation. The Stratonovich integral considered as a function of the upper limit of integration is not a martingale, as is the case for the Ito integral, and the simple expressions for the moments of the Ito integral (Theorem 2.5) do not hold for the Stratonovich integral. That the Stratonovich approach does not transform the probability law of the input Brownian motion process $W(t)$ as neatly as does the Ito, in this sense, seems to be the price of preserving the usual rules for its stochastic calculus.

This section is concluded with the statements of the corresponding multidimensional results now. Let X be a solution of (4.1) where f is an n-vector-valued function, G is an $n \times m$-matrix-valued function, and W is an m-dimensional Wiener process with independent components W_j, $1 \leq j \leq m$. Let

$$H(t,x) = \mathrm{col}[h_1(t,x), \ldots, h_m(t,x)], \qquad t \in [a,b], x \in R^n$$

be an $n \times m$-matrix-valued function with columns $h_j(t,x)$ (n-vector-valued functions), and their partial derivatives $\partial h_j/\partial x_i$ continuous in both variables; also assume, analogously to (4.4),

$$\int_a^b E\|H(t, X(t))\|^2 \, dt < \infty. \qquad (4.27)$$

(Here $\|\ \|$ denotes the matrix norm compatible with whatever norm is being used in R^n.)

The family of stochastic integrals, indexed by the parameter λ, then can be defined by the n-vector equation

$$\int_a^b H(t, X(t)) \, dW(t)$$
$$(\lambda)$$

$$= \lim_{\delta \to 0} \sum_{i=1}^{N-1} H(t_i, \lambda X(t_{i+1}) + (1-\lambda)X(t_i)) \, \Delta W_i \quad (4.28)$$

where $a = t_0 < t_1 < \cdots < t_N = b$ is a δ-partition of the interval $[a,b]$, $\Delta W_i = W(t_{i+1}) - W(t_i)$, $\lambda \in [0,1]$, and the limit in (4.28) is taken in the mean square sense. In terms of the columns of H, (4.28) may be written

$$
\int_{a\ (\lambda)}^{b} \sum_{j=1}^{m} h_j(t, X(t))\, dW_j(t)
$$

$$
= \lim_{\delta \to 0} \sum_{i=1}^{N-1} \left(\sum_{j=1}^{m} h_j(t, \lambda X(t_{i+1}) + (1-\lambda)X(t_i))\, \Delta W_{ji} \right) \quad (4.29)
$$

where $\Delta W_{ji} = W_j(t_{i+1}) - W_j(t_i)$, and the conclusion of Theorem 3.13 takes the form

$$
\int_{a\ (\lambda)}^{b} \sum_{j=1}^{m} h_j\, dW_j = \int_{a}^{b} \sum_{j=1}^{m} h_j\, dW_j + \lambda \int_{a}^{b} \sum_{j=1}^{m} \sum_{i=1}^{n} \frac{\partial h_j}{\partial x_i} g_{ij}\, dt. \quad (4.30)
$$

Finally, when G satisfies the conditions given above on H, one can compare the solutions of the stochastic differential equation (4.1) under different interpretations of the stochastic integral

$$
\int G(t, X(t))\, dW(t).
$$

In particular, for $\lambda = 1/2$, one has, analogously to (4.22) to (4.24), that the Stratonovich version of

$$
dX(t) = f(t, X(t))\, dt + \sum_{j=1}^{m} g_j(t, X(t))\, dW_j(t) \quad (4.31)
$$

is equivalent to the Ito equation

$$
dX(t) = \left\{ f(t, X(t)) + \frac{1}{2} \sum_{i=1}^{n} \sum_{j=1}^{m} \left[\frac{\partial g_j}{\partial x_i} g_{ij} \right](t, X(t)) \right\} dt
$$

$$
+ \sum_{j=1}^{m} g_j(t, X(t))\, dW_j(t) \quad (4.32)
$$

whereas the Ito interpretation of (4.31) corresponds to the Stratonovich equation

$$
d_S X(t) = \left\{ f(t, X(t)) - \frac{1}{2} \sum_{i=1}^{n} \sum_{j=1}^{m} \left[\frac{\partial g_j}{\partial x_i} g_{ij} \right](t, X(t)) \right\} dt
$$

$$
+ \sum_{j=1}^{m} g_j(t, X(t))\, dW_j(t). \quad (4.33)
$$

3.5 SOLUTIONS OF SOME PARTIAL DIFFERENTIAL EQUATIONS: EXIT PROBABILITIES

It was mentioned in Section 1.8 that the solution $u(t, x)$ of Kolmogorov's equation

$$Lu = u_t + f(t, x) \cdot u_x + \tfrac{1}{2} \operatorname{tr}(G(t, x) G(t, x)^T u_{xx}) = 0 \tag{5.1}$$

subject to the condition

$$u(t, x) \longrightarrow h(x) \qquad \text{as } t \uparrow T \tag{5.2}$$

can be written in the form

$$u(t, x) = E_{t,x} h(X(t)) \tag{5.3}$$

if h is a bounded measurable function, and $X(t)$ is an n-dimensional diffusion process whose drift f and diffusion GG^T coefficients satisfy certain regularity assumptions. In particular for $A \in \mathcal{B}^n$, the Borel sets in R^n, and $h(x) = I_A(x)$, the indicator function corresponding to A, the solution of (5.1) and (5.2) is

$$E_{t,x} I_A(X(T)) = P(X(T) \in A \mid X(t) = x),$$

namely, the conditional probability that the process X is in A at time T given that it had value x at a previous time t. Assuming the existence and uniqueness of solutions, Ito's formula facilitates showing that solutions of similar initial boundary value problems involving the operator L can be represented as conditional probabilities of the associated diffusion process. Chapter 6 in Friedman (1975) contains a number of these results. In this section one such basic result is given which permits representation of exit statistics for a diffusion process from a bounded region in terms of the solution of a Dirichlet problem.

Specifically, let D be a bounded domain in R^n with a smooth boundary ∂D and consider the Dirichlet problem

$$Mu = f(x) \cdot u_x + \tfrac{1}{2} \operatorname{tr}(G(x) G(x)^T u_{xx}) = k(x), \qquad x \in D \tag{5.4}$$

$$u(x) = h(x), \qquad x \in \partial D \tag{5.5}$$

where the n-vector functions f, k, and h, and the $n \times m$-matrix function G satisfy the following conditions.

(a) M is *uniformly elliptic* in D, that is, writing $GG^T = B = \{b_{ij}\}$, there is a positive number μ such that

$$\sum_{i,j=1}^{n} b_{ij}(x) \xi_i \xi_j \geq \mu |\xi|^2$$

if $x \in D$, $\xi = \{\xi_i\} \in R^n$.

(b)   The functions f, B, and k are uniformly Lipschitz continuous on $\overline{D}$, and h is continuous on ∂D.

Conditions a and b assure the existence of a unique solution of the Dirichlet problem (5.4) and (5.5).

THEOREM 3.14   Let $W(t) = (W_1(t), \ldots, W_m(t))^T$ denote a standard m-dimensional Wiener process, and let $X(t)$ be a solution of the stochastic differential equation

$$dX(t) = f(X(t))dt + G(X(t))\, dW(t), \qquad (5.6)$$

where the functions f and G satisfy conditions a and b above. If τ denotes the first exit time of X from D, τ is finite w.p. 1, and the solution $u(x)$ of the Dirichlet problem (5.4), (5.5) can be written

$$u(x) = E_x \left[h(X(\tau)) - \int_0^\tau k(X(s))\, ds \right] \qquad (5.7)$$

for any $x \in D$, where E_x denotes the conditional expectation given $X(0) = x$.

Proof: Fix $x \in D$, and let $X(t)$ denote the solution of (5.6) satisfying $X(0) = x$. Since Ito's formula is valid for functions of the solution process on bounded Markov time intervals, writing $\tau(t) = \tau \wedge t$,

$$u(X(\tau(t))) = u(x) + \int_0^{\tau(t)} Mu(X(s))\, ds$$

$$+ \int_0^{\tau(t)} u_x^T G(X(s))\, dW(s), \quad (5.8)$$

for any twice continuously differentiable function u and any $t > 0$. Taking expected value, and assuming that u satisfies (5.4), from (5.8) one obtains

$$Eu(X(\tau(t))) = u(x) + E \int_0^{\tau(t)} k(X(s))\, ds. \qquad (5.9)$$

If $X(t)$ represents any solution of (5.6), (5.9) becomes

$$E_x u(X(\tau(t))) = u(x) + E_x \int_0^{\tau(t)} k(X(s))\, ds. \qquad (5.10)$$

Taking $k \equiv 1$, and noting that u is bounded on $\overline{D}$, (5.10) yields the estimate for all t

$$E_x(\tau(t)) \le C \qquad (5.11)$$

if $|u(x)| \leq C/2$, $x \in \overline{D}$. Letting $t \to \infty$ in (5.11) gives $E_x \tau < \infty$, which means that solutions of (5.6) originating in D must exit D in finite time w.p. 1. Letting $t \to \infty$ in (5.10), then, gives the desired result, equation (5.7). □

EXAMPLE 3.4 Take $k \equiv -1$ and $h \equiv 0$. Equation (5.7) becomes

$$u(x) = E_x \tau. \tag{5.12}$$

The mean first exit time of the process conditioned on starting at the point x is the solution of the corresponding Dirichlet problem.

EXAMPLE 3.5 Let $k \equiv 0$, and let h be a continuous approximation of I_A, the indicator function of a Borel subset $A \subseteq \partial D$. (If A is a connected component of ∂D, then h can be taken to be I_A.) The solution $u(x)$ of (5.4), (5.5), in this case, has the form

$$u(x) = E_x h(X(\tau)), \tag{5.13}$$

which approximates the conditional probability

$$P(X(\tau) \in A \mid X(0) = x) \tag{5.14}$$

that first exit from D occurs through the portion A of the boundary.

EXAMPLE 3.6 Consider the scalar equation

$$dX = f(X)\,dt + g(X)\,dW, \tag{5.15}$$

with the domain $D = (\alpha, \beta)$, some constants $\alpha < \beta$; if f and g are continuously differentiable on $[\alpha, \beta]$, and $g(x) \neq 0$, $x \in [\alpha, \beta]$, conditions (a) and (b) hold.

Following Example 3.4, if $\tau = \tau_{[\alpha,\beta]}$, the exit time from the interval $[\alpha, \beta]$ for the solution $X(t)$ of (5.15), then the expected exit time

$$u(x) = E_x \tau$$

solves the boundary value problem

$$\begin{aligned} f(x)u'(x) + \tfrac{1}{2}g^2(x)u''(x) &= -1, \qquad x \in (\alpha, \beta) \\ u(\alpha) &= 0, \qquad u(\beta) = 0. \end{aligned} \tag{5.16}$$

From Example 3.5, taking $A = \{\beta\}$, the probability of exit through the boundary point β

$$u(x) = P_x(X(\tau) = \beta)$$

solves the boundary value problem

$$f(x)u'(x) + \tfrac{1}{2}g^2(x)u''(x) = 0$$
$$u(\alpha) = 0, \qquad u(\beta) = 1. \tag{5.17}$$

The solutions of these two scalar boundary value problems can be computed in terms of the functions

$$\varphi(x) = \exp\left\{-\int_\alpha^x \frac{2f(y)}{g^2(y)}\, dy\right\}$$

and

$$\xi(x) = \varphi(x) \int_\alpha^x 2[g^2(y)\varphi(y)]^{-1}\, dy.$$

Indeed letting

$$\psi(x) = \int_\alpha^x \varphi(y)\, dy \qquad \text{and} \qquad \theta(x) = \int_\alpha^x \xi(y)\, dy,$$

the solution of (5.17) can be written

$$u(x) = \psi(x)/\psi(\beta) \tag{5.18}$$

and the solution of (5.16) is given by

$$u(x) = [\psi(x)\theta(\beta) - \psi(\beta)\theta(x)]/\psi(\beta). \tag{5.19}$$

The examples in this section suggest using partial differential equation theory, for example, the maximum principle, to obtain estimates for the distribution of exit points and the mean exit time.   Stochastic analogues of other versions of the comparison principle for deterministic differential equations can be used to study further qualitative properties of the solution process.   In Chapter 5, it is shown that the properties of stochastic boundedness, stochastic stability, and pathwise uniqueness for solutions of stochastic differential equations can be obtained by constructing an auxiliary function V of the solution. A stochastic version of the simple inequality $\dot{V} \le 0$ yields the key property that V applied to a solution is a supermartingale.   But first, the question of which stochastic differential equations can be explicitly solved is addressed.

EXERCISES

3.1 Write out the solution $x(t)$ of the initial value problem

$$\frac{dx}{dt} = E(Y_1(t))x + E(Y_2(t))$$

$$x(0) = EX_0,$$

on the interval $[0, T]$, assuming that $Y_1(t)$ and $Y_2(t)$ are continuous w.p. 1 processes with finite means on $[0, T]$; compare with (0.7).

3.2 Prove Lemma 3.2 (Bellman-Gronwall inequality). [Hint: Let $c(t) = b(t) + L \int_0^t e^{L(t-s)} b(s)\, ds - a(t)$ and verify that $c(t) \geq L^{n+1}/n! \int_0^t (t-s)^n c(s)\, ds$ for arbitrary positive integers n.]

3.3 Show that if $f(t, 0)$ is bounded and f satisfies the Lipschitz condition: for some constant K,

$$|f(t, x) - f(t, y)| \leq K|x - y|,$$

then f also satisfies the linear growth condition:

$$|f(t, x)|^2 \leq L^2(1 + |x|^2), \qquad \text{where } L \text{ is a constant.}$$

3.4 Under the assumptions of Theorem 3.5 verify (1.27).

3.5 Compute the stochastic differential $dF(X(t))$ if $F(x) = \sin x$ and $X(t)$ is a solution of the equation

$$dX = \sec X \left[\tfrac{1}{2}(1 + \sec X \tan X)\, dt + dW \right].$$

Conclude that $X(t) = \sin^{-1}[\sin X(t_0) + \tfrac{1}{2}(t - t_0) + W(t) - W(t_0)]$.

3.6 If $X(t)$ is a solution of the vector stochastic equation (0.10), show that

$$|X(t)|^{2n} = |X(0)|^{2n} + \int_0^t 2n|X(s)|^{2n-2} X(s)^T f(s, X(s))\, ds$$

$$+ \int_0^t n|X(s)|^{2n-2} |G(s, X(s))|^2\, ds$$

$$+ \int_0^t 2n(n-1)|X(s)|^{2n-4} |X(s)^T G(s, X(s))|^2\, ds$$

$$+ \int_0^t 2n|X(s)|^{2n-2} X(s)^T G(s, X(s))\, dW(s).$$

3.7 Prove (3.13).

3.8 Prove (3.32) and (3.33).

3.9   Extend Example 3.3 to the case where h is a function of t and x, $\partial h/\partial t$ and $\partial^2 h/\partial x^2$ are continuous, and $X(t)$ is the solution of the Stratonovich-interpreted equation (4.16) with continuous $\partial g/\partial x$.

3.10 Verify that (5.18) solves the boundary value problem (5.17) and that (5.19) solves (5.16).

4

Solving Stochastic
Differential Equations

4.0 INTRODUCTION: REDUCING STOCHASTIC DIFFERENTIAL EQUATIONS

Having discussed existence and uniqueness of solutions among other basic properties, it is appropriate now to consider the question of which stochastic differential equations can be solved by quadratures. When can solutions be represented explicitly in terms of ordinary and stochastic integrals?

An approach to answering this question is supplied by the notion of reducibility which can be implemented by a change of variables in the stochastic equation. A smooth function $h(t, x)$ having an inverse $k(t, x)$,

$$h(t, k(t, x)) = x, \qquad k(t, h(t, x)) = x, \tag{0.1}$$

affects a change of variables in the stochastic differential equation

$$dX(t) = f(t, X(t))\, dt + G(t, X(t))\, dW(t) \tag{0.2}$$

according to Ito's formula. Indeed, the process

$$Y(t) = h(t, X(t))$$

satisfies the equation

$$dY(t) = \bar{f}(t, Y(t))\, dt + \bar{G}(t, Y(t))\, dW(t) \tag{0.3}$$

where

$$\bar{f}(t, y) = \left[h_t + h_x f + \tfrac{1}{2}\operatorname{tr}(GG^T h_{xx}) \right](t, k(t, y)) \tag{0.4}$$

109

and

$$\bar{G}(t,y) = [h_x G](t, k(t,y)).$$ (0.5)

Here the following notation is used: Writing $h = \{h_i\}$, h_t denotes the vector $\{\partial h_i / \partial t\}$; h_x is the matrix $\{\partial h_i / \partial x_j\}$; and $\mathrm{tr}(GG^T h_{xx})$ stands for the vector

$$\left\{ \sum_{j,k=1}^{n} \sum_{\nu=1}^{n} g_{j\nu} g_{k\nu} \frac{\partial^2 h_i}{\partial x_j \, \partial x_k} \right\}.$$

Equation (0.2) is said to be *reducible* if such a function h can be found so that the functions $\bar{f}$ and $\bar{G}$ given by (0.4) and (0.5) respectively are independent of y. In this case, the change of variables $y = h(t,x)$ permits the explicit representation of the solution $X(t)$ of (0.2) as

$$X(t) = k(t, Y(t))$$ (0.6)

$$Y(t) = h(0, X(0)) + \int_0^t \bar{f}(s)\,ds + \int_0^t \bar{G}(s)\,dW(s).$$ (0.7)

For example, if an invertible function h can be determined such that $\bar{f} = c_1$ and $\bar{G} = C_2$ (c_1 is a constant vector and C_2 a constant matrix), then (0.6) and (0.7) yield

$$X(t) = k(t, h(0, X(0)) + c_1 t + C_2 W(t)).$$

In the following section a condition on f and G characterizing reducibility is obtained for the scalar case. The explicit solution of the linear scalar equation is computed even though this equation is not reducible in general. Section 4.2 deals with linear systems of stochastic differential equations. Section 4.3 concludes this chapter with a presentation of a more general concept than solvability by quadratures—namely, simulation of the solution by a system that separates the intrinsically stochastic portion from the deterministic part of the equation.

4.1   SCALAR DIFFERENTIAL EQUATIONS

The equation

$$dX(t) = f(t, X(t))\,dt + g(t, X(t))\,dW(t)$$ (1.1)

where f and g are real-valued functions and $W(t)$ is a Wiener process is reducible if (from (0.4) and (0.5))

$$\left[\frac{\partial h}{\partial t} + \frac{\partial h}{\partial x} f + \frac{1}{2} \frac{\partial^2 h}{\partial x^2} g^2 \right](t, x) = \bar{f}(t)$$ (1.2)

and

$$\left[\frac{\partial h}{\partial x}g\right](t,x) = \bar{g}(t) \tag{1.3}$$

for some smooth invertible function h. To see what specific requirements (1.2) and (1.3) impose on the functions f and g, one proceeds as follows, under the assumptions that $g \neq 0$ and f and g possess the indicated derivatives.

Differentiating (1.2) with respect to x gives

$$\frac{\partial^2 h}{\partial x\,\partial t} + \frac{\partial}{\partial x}\left\{\frac{\partial h}{\partial x}f + \frac{1}{2}\frac{\partial^2 h}{\partial x^2}g^2\right\} = 0. \tag{1.4}$$

From (1.3) one obtains

$$\frac{\partial h}{\partial x}(t,x) = \frac{\bar{g}(t)}{g(t,x)} \tag{1.5}$$

from which it follows that

$$\frac{\partial^2 h}{\partial t\,\partial x} = \frac{g(t,x)\bar{g}'(t) - \bar{g}(t)[\partial g/\partial t](t,x)}{g^2(t,x)} \tag{1.6}$$

and

$$\frac{\partial^2 h}{\partial x^2} = \frac{-\bar{g}(t)[\partial g/\partial x](t,x)}{g^2(t,x)}. \tag{1.7}$$

Substituting (1.5), (1.6), and (1.7) into (1.4) gives

$$\frac{\bar{g}'}{g} - \bar{g}\left\{\frac{\partial g/\partial t}{g^2} - \frac{\partial}{\partial x}\left(\frac{f}{g}\right) + \frac{1}{2}\frac{\partial^2 g}{\partial x^2}\right\} = 0$$

or

$$\bar{g}' = \bar{g}g\left\{\frac{1}{g^2}\frac{\partial g}{\partial t} - \frac{\partial}{\partial x}\left(\frac{f}{g}\right) + \frac{1}{2}\frac{\partial^2 g}{\partial x^2}\right\}. \tag{1.8}$$

Since the left side of (1.8) is independent of x, it follows that

$$\frac{\partial}{\partial x}\left[g\left\{\frac{1}{g^2}\frac{\partial g}{\partial t} - \frac{\partial}{\partial x}\left(\frac{f}{g}\right) + \frac{1}{2}\frac{\partial^2 g}{\partial x^2}\right\}\right] = 0. \tag{1.9}$$

If (1.9) holds, $\bar{g} \neq 0$ can be determined from (1.8), and h (which is at least locally invertible, since $\partial h/\partial x \neq 0$) can be computed from (1.5). Since (1.8) and (1.4) are equivalent, the function $\bar{f}$ obtained in this case by (1.2) is indeed independent of x. Thus the following result is proved.

THEOREM 4.1   Equation (1.1) is reducible if and only if the coefficient functions f and g satisfy (1.9).

EXAMPLE 4.1   Autonomous equations ($f(t,x) = f(x)$, $g(t,x) = g(x)$) are reducible if and only if, for some constant c,

$$g\left[\tfrac{1}{2}g'' - \left(\frac{f}{g}\right)'\right] = c. \qquad (1.10)$$

Here (1.8) implies that one can take $\bar{g}(t) = e^{ct}$, and consequently (1.5) gives

$$h(t,x) = e^{ct}\int_a^x \frac{dy}{g(y)},$$

for arbitrary a. Then $\bar{f}(t)$ is calculated from (1.2). For the special case $f(x) \equiv 0$ in this example, (1.10) reduces to

$$\tfrac{1}{2}gg'' = c.$$

If $c \neq 0$, letting $v = g'$ gives the first-order equation

$$gv\frac{dv}{dg} = 2c.$$

For g positive on some interval about $x = 0$, the solution of this equation is

$$v = \pm\sqrt{4c\ln g + v_1^2},$$

where v_1 is an arbitrary constant; integrating again yields

$$x = \pm\int_{g(0)}^{g(x)} \frac{d\gamma}{\sqrt{4c\ln\gamma + v_1^2}};$$

that is, in this case, for reducibility, $g(x)$ must be given implicitly by the last equation. If $c = 0$, then g is linear, and hence the equation

$$dX(t) = [\lambda X(t) + \mu]\,dW(t)$$

for constants λ and μ is reducible. In particular, one can take $\bar{g}(t) \equiv 1$ and

$$h(t,x) = h(x) = \int_{x_0}^x \frac{dy}{\lambda y + \mu} = \frac{1}{\lambda}\ln\left|\frac{\lambda x + \mu}{\lambda x_0 + \mu}\right|.$$

Then $h''(x) = -\lambda/(\lambda x + \mu)^2$, and from (1.2) one obtains

$$\bar{f}(t) = \tfrac{1}{2}h''(x)g^2(x) = \frac{1}{2}\frac{-\lambda}{(\lambda x + \mu)^2}(\lambda x + \mu)^2 \equiv -\frac{\lambda}{2}.$$

The reduced equation, therefore, is

$$dY(t) = -\frac{\lambda}{2} dt + dW(t),$$

which gives

$$Y(t) = Y(0) - \frac{\lambda}{2}t + W(t).$$

Since $Y(t) = h[X(t)]$, finally one has

$$\frac{1}{\lambda} \ln \left| \frac{\lambda X(t) + \mu}{\lambda X(0) + \mu} \right| = -\frac{\lambda}{2}t + W(t)$$

or

$$X(t) = \left[X(0) + \frac{\mu}{\lambda} \right] \exp \left\{ -\frac{\lambda^2}{2}t + \lambda W(t) \right\} - \frac{\mu}{\lambda}.$$

In general, linear equations are not reducible. Writing

$$f(t,x) = f_1(t) + f_2(t)x \qquad \text{and} \qquad g(t,x) = g_1(t) + g_2(t)x,$$

the reducibility condition (1.9) becomes

$$g_1 g_2' - (f_1 g_2 - f_2 g_1 + g_1')g_2 = 0.$$

This equation implies that homogeneous $(f_1(t) = g_1(t) \equiv 0)$ equations and narrow-sense linear $(g_2(t) \equiv 0)$ equations are reducible.

Despite the fact that nonhomogeneous linear equations are not reducible in general, explicit solutions can be obtained, as in the ordinary differential equations case, by the variation of parameters technique. First the solution of the homogeneous equation

$$dX(t) = f_2(t)X(t)\, dt + g_2(t)X(t)\, dW(t), \tag{1.11}$$

which is reducible, is obtained. Assuming, without loss of generality, that $X(0) > 0$, equations (1.8), (1.5), and (1.2) lead to the choices $\bar{g}(t) = g_2(t)$, $h(x) = \ln x$, and $\bar{f}(t) = f_2(t) - \frac{1}{2}g_2^2(t)$. The solutions of (1.11) as then determined from (0.6) and (0.7) is

$$X(t) = X(0) \exp \left\{ \int_0^t [f_2(s) - \tfrac{1}{2}g_2^2(s)]\, ds + \int_0^t g_2(s)\, dW(s) \right\}. \tag{1.12}$$

Denote by $X_0(t)$, the solution (1.12) with $X(0) = 1$. Now, writing the solution of the nonhomogeneous equation

$$dX(t) = [f_1(t) + f_2(t)X(t)]\, dt + [g_1(t) + g_2(t)X(t)]\, dW(t) \tag{1.13}$$

as $X(t) = Y(t)X_0(t)$, the problem is to determine $Y(t)$.

$$Y(t) = X_0^{-1}(t)X(t), \qquad (1.14)$$

where

$$X_0^{-1}(t) = \exp\left\{-\int_0^t [f_2(s) - \tfrac{1}{2}g_2^2(s)]\,ds - \int_0^t g_2(s)\,dW(s)\right\}; \qquad (1.15)$$

applying the preceding result for linear homogeneous equations yields

$$dX_0^{-1}(t) = X_0^{-1}(t)\left\{[-f_2(t) + g_2^2(t)]\,dt - g_2(t)\,dW(t)\right\}. \qquad (1.16)$$

The stochastic differential of Y can be computed using Example 2.1 and (1.13) and (1.16) as

$$
\begin{aligned}
dY(t) &= X_0^{-1}(t)\,dX(t) + X(t)\,dX_0^{-1}(t) - [g_1(t) + g_2(t)X(t)]X_0^{-1}(t)g_2(t)\,dt \\
&= X_0^{-1}(t)\left\{[f_1(t) + f_2(t)X(t)]\,dt + [g_1(t) + g_2(t)X(t)]\,dW(t)\right\} \\
&\quad + X(t)X_0^{-1}(t)\left\{[-f_2(t) + g_2^2(t)]\,dt - g_2(t)\,dW(t)\right\} \\
&\quad - [g_1(t) + g_2(t)X(t)]X_0^{-1}(t)g_2(t)\,dt \\
&= X_0^{-1}(t)\left\{[f_1(t) - g_1(t)g_2(t)]\,dt + g_1(t)\,dW(t)\right\}.
\end{aligned}
$$

Therefore,

$$
\begin{aligned}
Y(t) = Y(0) &+ \int_0^t X_0^{-1}(s)[f_1(s) - g_1(s)g_2(s)]\,ds \\
&+ \int_0^t X_0^{-1}(s)g_1(s)\,dW(s), \qquad (1.17)
\end{aligned}
$$

which establishes the following result.

THEOREM 4.2   The solution $X(t)$ of the nonhomogeneous linear stochastic differential equation (1.13) can be written

$$
\begin{aligned}
X(t) = X_0(t)\Bigg\{ X(0) &+ \int_0^t X_0^{-1}(s)[f_1(s) - g_1(s)g_2(s)]\,ds \\
&+ \int_0^t X_0^{-1}(s)g_1(s)\,dW(s) \Bigg\} \qquad (1.18)
\end{aligned}
$$

where

$$X_0(t) = \exp\left\{\int_0^t [f_2(s) - \tfrac{1}{2}g_2^2(s)]\,ds + \int_0^t g_2(s)\,dW(s)\right\}.$$

EXAMPLE 4.2 The Langevin equation

$$\dot{X}(t) = -\alpha X(t) + \sigma \mathcal{N}(t)$$

where $\mathcal{N}(t)$ is a white noise process, describes a scalar component of the velocity of the Brownian motion of a particle. The corresponding stochastic differential equation

$$dX(t) = -\alpha X(t)\, dt + \sigma\, dW(t)$$

is narrow-sense linear and autonomous. According to (1.18) the solution is

$$X(t) = e^{-\alpha t}\left[X(0) + \int_0^t e^{\alpha s}\sigma\, dW(s)\right]. \tag{1.19}$$

The process $X(t)$ given by (1.19) is known as the Ornstein-Uhlenbeck process.

Theorem 4.2 motivates the consideration of a more general type of reduction than that indicated by (1.2) and (1.3), namely, the reduction to a general linear equation. For autonomous equations, this requires [from the scalar case of (0.4) and (0.5)]

$$\left[fh' + \tfrac{1}{2}g^2 h''\right](k(y)) = \alpha + \beta y \tag{1.20}$$

and

$$gh'(k(y)) = \gamma + \delta y \tag{1.21}$$

for some constants α, β, γ, and δ. As in the case of reducibility, one can seek specific conditions on f and g which make (1.20) and (1.21) possible.
Indeed, for $y = h(x)$ where h and k are inverses, one has

$$f(x)h'(x) + \tfrac{1}{2}g^2(x)h''(x) = \alpha + \beta h(x) \tag{1.22}$$

and

$$g(x)h'(x) = \gamma + \delta h(x). \tag{1.23}$$

In case $\delta \neq 0$, assuming once again that $g \neq 0$, the function

$$h(x) = Ce^{\delta B(x)} - \gamma/\delta$$

satisfies (1.23), if $B(x) = \int_a^x 1/g(y)\, dy$ and C is an arbitrary constant. Substituting this expression for h into (1.22) gives

$$\left\{\frac{f(x)}{g(x)}\delta + \tfrac{1}{2}g^2(x)\left[\frac{\delta^2}{g^2(x)} - \frac{\delta g'(x)}{g^2(x)}\right]\right\}Ce^{\delta B(x)} = Ce^{\delta B(x)} - \frac{\beta\delta}{\gamma} + \alpha.$$

It follows that

$$\left\{\left[\frac{f(x)}{g(x)} - \tfrac{1}{2}g'(x)\right]\delta + \tfrac{1}{2}\delta^2 - \beta\right\}e^{\delta B(x)} = \frac{\alpha\gamma - \beta\delta}{C\gamma}. \tag{1.24}$$

Letting

$$A(x) = \frac{f(x)}{g(x)} - \tfrac{1}{2}g'(x)$$

in (1.24), and differentiating results in

$$\left[\{ A(x)\delta + \tfrac{1}{2}\delta^2 - \beta \} / g(x) + A'(x) \right] \delta e^{\delta B(x)} = 0$$

and, differentiating again, after multiplying by $[g(x)/\delta]e^{-\delta B(x)}$, leads to

$$\delta A'(x) + (gA')'(x) = 0. \tag{1.25}$$

One obtains from (1.25) that either

$$A'(x) = 0 \quad \text{or} \quad \left(\frac{[gA']'}{A'} \right)'(x) = 0. \tag{1.26}$$

Conversely, if the latter part of (1.26) is satisfied, it is easy to verify that the choices

$$h(x) = Ce^{\delta B(x)}, \quad \text{where } \delta = -\frac{[gA']'(x)}{A'(x)}$$

and C is some constant, reduce the original equation to linear form. For the $\delta = 0$ case, the choice $h(x) = \gamma B(x) + C$ similarly leads to the reducibility condition

$$[gA']'(x) = 0. \tag{1.27}$$

Note that (1.27) is precisely (1.9) in the autonomous case. Since (1.27) implies (1.26), equation (1.26) constitutes the sought-after reducibility criterion. The preceding discussion is summarized in the next theorem.

THEOREM 4.3   The autonomous scalar stochastic differential equation

$$dX(t) = f(X(t))\,dt + g(X(t))\,dW(t)$$

is reducible to a scalar linear stochastic differential equation if and only if (1.26) holds.

EXAMPLE 4.3   The scalar stochastic differential equation

$$dX = \alpha X(1 - (X/K))\,dt + \beta X\,dW \tag{1.28}$$

satisfies (1.26). The method of this section leads to the explicit solution

$$X(t) = \frac{\exp\left\{ (\alpha - \beta^2/2)t + \beta W(t) \right\}}{(X(0))^{-1} + (\alpha/K)\int_0^t \exp\left\{ (\alpha - \beta^2/2)s + \beta W(s) \right\} ds}. \tag{1.29}$$

4.2   LINEAR STOCHASTIC DIFFERENTIAL EQUATIONS

In this section the general linear case of equation

$$dX(t) = [f(t) + F(t)X(t)]\, dt + \sum_{i=1}^{m} [g_i(t) + G_i(t)X(t)]\, dW_i(t) \qquad (2.1)$$

is considered; here $X(t)$ represents a random n-vector process, $W(t) = [W_1(t),\ldots,W_m(t)]^T$ is an m-vector standard Wiener process, $F(t)$ and the $G_i(t)$ are $n \times n$-matrix functions, and $f(t)$ and the $g_i(t)$ are n-vector functions, respectively. By Theorem 3.1, for any initial value X_0 independent of $W(t)$ on $[0,T]$, there exists a unique solution $X(t)$ of (2.1) satisfying $X(0) = X_0$ if the functions $f(t)$, $F_i(t)$, $g_i(t)$, and $G_i(t)$ are bounded and measurable on $[0,T]$. The following result extends Theorem 4.2 of the previous section to cover the preceding linear systems case and is totally analogous to the ordinary differential equations situation. The proof is similar to that for Theorem 4.2.

THEOREM 4.4   Let $\Phi(t)$ be the fundamental matrix of the corresponding [to (2.1)] homogeneous equation

$$d\Phi(t) = F(t)\Phi(t)\, dt + \sum_{i=1}^{m} G_i(t)\Phi(t)\, dW_i(t), \qquad (2.2)$$

that is, $\Phi(t)$ is the $n \times n$-matrix solution of (2.2) which satisfies $\Phi(0) = I$ ($n \times n$ identity matrix). The solution of (2.1) can be written

$$X(t) = \Phi(t) \left\{ X(0) + \int_0^t \Phi^{-1}(s) \left[f(s) - \sum_{i=1}^{m} G_i(s)g_i(s) \right] ds \right.$$

$$\left. + \int_0^t \Phi^{-1}(s) \sum_{i=1}^{m} g_i(s)\, dW_i(s) \right\}. \qquad (2.3)$$

The analogy with the scalar case Theorem 4.2 breaks down (just as it does in the ordinary differential equations case) in that an explicit expression for $\Phi(t)$ cannot be given in general, even when the matrices F and G_i are constant. However, if, in addition, the matrices F and G_i pairwise commute, that is, $FG_i = G_iF$, $G_iG_j = G_jG_i$ all i,j, then

$$\Phi(t) = \exp \left\{ \left(F - \frac{1}{2}\sum_{i=1}^{m} G_i^2 \right) t + \sum_{i=1}^{m} G_iW_i(t) \right\} \qquad (2.4)$$

can be established by Ito's formula, as the expression for $X_0(t)$ was derived in Theorem 4.2.

In the narrow-sense linear case $[G_i(t) \equiv 0,\ \text{all } i]$, (2.3) has the simplified form

$$X(t) = \Phi(t)$$
$$\times \left\{ X(0) + \int_0^t \Phi^{-1}(s) f(s)\, ds + \int_0^t \Phi^{-1}(s) \sum_{i=1}^m g_i(s)\, dW_i(s) \right\} \quad (2.5)$$

and the fundamental matrix $\Phi(t)$ corresponds to the solution of the deterministic initial value problem

$$\frac{d\Phi}{dt} = F(t) \qquad \Phi(0) = I\ (n \times n \text{ identity matrix}),$$

that is, $\Phi(t)$ is the fundamental matrix for the deterministic part of (2.1).

EXAMPLE 4.4   The second-order Langevin equation

$$\ddot{X}(t) = K(t, X(t), \dot{X}(t)) + \sigma \mathcal{N}(t)$$

describes the motion of a particle in a noise-perturbed force field and can be interpreted as the vector stochastic equation

$$d \begin{bmatrix} X(t) \\ V(t) \end{bmatrix} = \begin{bmatrix} V(t) \\ K(t, X(t), V(t)) \end{bmatrix} dt + \begin{bmatrix} 0 \\ \sigma \end{bmatrix} dW(t) \quad (2.6)$$

where $V(t) = \dot{X}(t)$. [See Chandrasekhar (1943), for example.] For the case of the harmonic oscillator, $K(t, x, \dot{x}) = -\nu^2 x - \beta \dot{x}$, (2.6) is the narrow-sense linear equation

$$d \begin{bmatrix} X(t) \\ V(t) \end{bmatrix} = F \begin{bmatrix} X(t) \\ V(t) \end{bmatrix} dt + \begin{bmatrix} 0 \\ \sigma \end{bmatrix} dW(t) \quad (2.7)$$

where

$$F = \begin{bmatrix} 0 & 1 \\ -\nu^2 & -\beta \end{bmatrix}.$$

In this case the fundamental matrix can be written as the matrix exponential

$$\Phi(t) = \exp Ft,$$

and so the solution of (2.7) can be written explicitly, via (2.5), as

$$\begin{bmatrix} X(t) \\ V(t) \end{bmatrix} = (\exp Ft) \left\{ \begin{bmatrix} X(0) \\ V(0) \end{bmatrix} + \int_0^t (\exp -Fs) \begin{bmatrix} 0 \\ \sigma \end{bmatrix} dW(s) \right\}. \quad (2.8)$$

Attention is now turned to the statistics of the solution process $X(t)$ of (2.1). The next result indicates that the first two moments of $X(t)$ can be obtained as solutions of initial value problems involving linear ordinary differential equations. Let

$$m(t) = EX(t) \quad \text{and} \quad M(t) = E(X(t)X(t)^T),$$

and assume $E|X(0)|^2 < \infty$.

THEOREM 4.5   The mean vector $m(t)$ is the unique solution of the initial value problem

$$\frac{dm}{dt} = f(t) + F(t)m, \qquad m(0) = EX(0) \tag{2.9}$$

and the second-moment matrix $M(t)$ is the unique solution of the initial value problem

$$\frac{dM}{dt} = F(t)M + MF(t)^T + \sum_{i=1}^{m} G_i(t)MG_i(t)^T$$

$$+ f(t)m(t)^T + m(t)f(t)^T$$

$$+ \sum_{i=1}^{m} [G_i(t)m(t)g_i(t)^T$$

$$+ g_i(t)m(t)^T G_i(t)^T + g_i(t)g_i(t)^T], \tag{2.10}$$

$$M(0) = E(X(0)X(0)^T).$$

Proof: Taking expected value in the integral form of equation (2.1) verifies the first assertion. Toward proving the second, recall that Ito's formula leads to the "product rule" for stochastic differentials, which, in this case, takes the form

$$d(X(t)X(t)^T) = X(t)\,dX(t)^T + (dX(t))X(t)^T$$

$$+ \sum_{i=1}^{m} (G_i(t)X(t) + g_i(t))(X(t)^T G_i(t)^T + g_i(t)^T)\,dt. \tag{2.11}$$

Substituting for $dX(t)$ and $dX(t)^T$ in (2.11), writing in integral form, and taking expected value leads to the integral equation equivalent to the initial value problem (2.10).   □

Expressions for other solution statistics, for example, the covariance and correlation matrices, can be derived starting from the explicit form

(2.3) of the solution. For the narrow-sense $[G_i(t) \equiv 0$, all $i]$ linear case, the fact that the fundamental matrix $\Phi(t)$ is not random implies that the solution is a Gaussian process if the initial value $X(0)$ is normally distributed, and allows a simplified expression for the correlations.

$$C_{XX}(s,t) = E(X(s)X(t)^T)$$

as follows. Setting

$$\mathcal{V}(t) = \int_0^t \Phi^{-1}(s) f(s)\, ds$$

and

$$\mathcal{W}(t) = \int_0^t \Phi^{-1}(s) \sum_{i=1}^m g_i(s)\, dW_i(s)$$

one obtains

$$C_{XX}(s,t) = \Phi(s) \left\{ M(0) + E(\mathcal{V}(s))m(0)^T + m(0)E(\mathcal{V}(t))^T \right.$$
$$\left. + C_{\mathcal{V}\mathcal{V}}(s,t) + C_{\mathcal{V}\mathcal{W}}(s,t) + C_{\mathcal{W}\mathcal{W}}(s,t) \right\} \Phi(t)^T. \qquad (2.12)$$

The correlation term $C_{\mathcal{W}\mathcal{W}}$ can be written as an ordinary integral by making use of the independence of stochastic integrals on disjoint intervals and Theorem 2.5 as follows. Setting $r = \min(s,t)$,

$$C_{\mathcal{W}\mathcal{W}}(s,t)$$

$$= E\left(\int_0^s \Phi^{-1}(u) \sum_{i=1}^m g_i(u)\, dW_i(u) \int_0^t \sum_{i=1}^m g_i(u)^T \Phi^{-1}(u)^T\, dW_i(u) \right)$$

$$= E\left(\int_0^r \Phi^{-1}(u) \sum_{i=1}^m g_i(u)\, dW_i(u) \int_0^r \sum_{i=1}^m g_i(u)^T \Phi^{-1}(u)^T\, dW_i(u) \right)$$

$$= \int_0^r E\left| \Phi^{-1}(u) \sum_{i=1}^m g_i(u) \right|^2 du = \int_0^r \left| \Phi^{-1}(u) \sum_{i=1}^m g_i(u) \right|^2 du.$$

For the narrow-sense linear equation of the form

$$dX(t) = F(t)X(t)\, dt + \sum_{i=1}^m g_i(t)\, dW_i(t) \qquad (2.13)$$

$\mathcal{V}(t) \equiv 0$, and so the correlation takes the form

$$C_{XX}(s,t) = \Phi(s) \left\{ M(0) + \int_0^{\min(s,t)} \left| \Phi^{-1}(u) \sum_{i=1}^m g_i(u) \right|^2 du \right\} \Phi(t)^T.$$

(2.14)

EXAMPLE 4.5   For the Ornstein-Uhlenbeck process (Example 4.2) with initial value $X(0)$ a normally distributed $\mathcal{N}(0, \sigma^2/2\alpha)$ random variable, it follows from (2.9) and (2.10) that $m(t) \equiv 0$ and $M(t) \equiv \sigma^2/2\alpha$. Since the stochastic equation

$$dX(t) = -\alpha X(t) \, dt + \sigma \, dW(t)$$

for the Ornstein-Uhlenbeck process $X(t)$ is of the form (2.13), the correlation can be obtained from (2.14):

$$C_{XX}(s,t) = e^{-\alpha s} \left\{ \frac{\sigma^2}{2\alpha} + \int_0^{\min(s,t)} e^{2\alpha u} \sigma^2 \, du \right\} e^{-\alpha t}$$

$$= \frac{e^{-\alpha |t-s|} \sigma^2}{2\alpha}.$$

(2.15)

Thus the solution $X(t)$ is a stationary Gaussian process having constant mean and variance. The form of (2.15) motivates the terminology "colored" Gaussian noise for the generalized process associated with the "derivative" of $X(t)$.

EXAMPLE 4.6   Returning to the stochastic harmonic oscillator [Example 4.4, equation (2.7)]

$$d \begin{bmatrix} X(t) \\ V(t) \end{bmatrix} = F \begin{bmatrix} X(t) \\ V(t) \end{bmatrix} dt + \begin{bmatrix} 0 \\ \sigma \end{bmatrix} dW(t)$$

with

$$F = \begin{bmatrix} 0 & 1 \\ -\nu^2 & -\beta \end{bmatrix}$$

one can obtain directly explicit expressions for the mean $EX(t)$ and correlation $E(X(s)X(t))$ of the position $X(t)$. Indeed, writing

$$m(t) = E \begin{bmatrix} X(t) \\ V(t) \end{bmatrix} \quad \text{and} \quad C(s,t) = E \begin{bmatrix} X(s) \\ V(s) \end{bmatrix} \begin{bmatrix} X(t) \\ V(t) \end{bmatrix}^T$$

the sought-after quantities are the first component of m and the first-row, first-column entry of C. From (2.9) and (2.14), recalling that the fundamental matrix here is of the form $\Phi(t) = \exp Ft$, one has

$$m(t) = m(0) \exp Ft$$

and   (2.16)

$$C(s,t) = (\exp Fs) \left[C(0,0) + \int_0^{\min(s,t)} \left| (\exp -Fu) \begin{pmatrix} 0 \\ \sigma \end{pmatrix} \right|^2 du \right] \exp F^T t.$$

4.3   SIMULATION BY A DECOUPLED SYSTEM

In Section 4.0 it was pointed out that the simple representation of the solution of a stochastic differential equation by quadratures, (0.6) and (0.7), followed from the reducibility of the equation. Conditions characterizing reducibility, however, are somewhat restrictive, even for scalar equations, as Theorems 4.1 and 4.3 in Section 4.1 illustrate. Suppose, in lieu of (0.6) and (0.7), one requires

$$X(t) = \varphi(t, Y(t), Z(t))$$ (3.1)

where Y and Z satisfy

$$dY(t) = \tilde{G}(t, Y(t)) \, d\tilde{W}(t)$$ (3.2)

$$dZ(t) = \tilde{f}(t, Y(t), Z(t)) \, dt$$ (3.3)

for some smooth functions φ, $\tilde{f}$, and $\tilde{G}$. If φ, $\tilde{f}$, $\tilde{G}$, and $\tilde{W}$ satisfying (3.1) to (3.3) exist, then the stochastic differential equation is said to be *simulated by the decoupled system* (3.2), (3.3). For (3.1) to (3.3) to be valid, the function φ, $\tilde{f}$, and $\tilde{G}$ must be related to f and G in (0.2) by

$$\left[\varphi_t + \varphi_z \tilde{f} + \tfrac{1}{2} \operatorname{tr}(\tilde{G}\tilde{G}^T \varphi_{yy}) \right](t, Y(t), Z(t)) = f(t, \varphi(t, Y(t), Z(t))$$

(3.4)

and

$$[\varphi_y \tilde{G}](t, Y(t), Z(t)) = G(t, \varphi(t, Y(t), Z(t)))$$ (3.5)

according to Ito's formula. [Here, as in Section 4.0, the notation is employed: for $\varphi = \{\varphi_i\}$, φ_t denotes the vector $\{\partial\varphi_i/\partial t\}$, φ_y and φ_z denote the matrices $\{\partial\varphi_i/\partial y_j\}$ and $\{\partial\varphi_i/\partial z_j\}$ respectively, and $\operatorname{tr}(\tilde{G}\tilde{G}^T \varphi_{yy})$ denotes the vector with components $\operatorname{tr}(\tilde{G}\tilde{G}^T \varphi_{iyy})$, where φ_{iyy} is the matrix

$\{\partial^2 \varphi_i/(\partial x_j\, \partial y_k)\}$.] The principal feature of the representation (3.1) to (3.3) is that it isolates the stochastic complexity of the original equation (0.2) in the simpler equation (3.2); for example, if (3.2) can be solved by quadratures, then the solution $X(t)$ of (0.2) is a function of the solution of the ordinary differential equation (3.3). The next theorem illustrates that simulation by a decoupled system generalizes the notion of solvability by quadratures.

THEOREM 4.6 If the stochastic differential equation (0.2) is reducible, then it can be simulated by a decoupled system.

Proof: Let $h(t,x)$, $k(t,x)$, $\bar{f}(t)$, and $\bar{G}(t)$ be functions satisfying (0.1), (0.4), and (0.5) and let

$$Y(t) = D(t)h(t, X(t)) \tag{3.6}$$

where

$$D(t) = \text{diag.}\left\{ \exp\left(-\int^t \bar{f}_1(s)\, ds \right), \ldots, \exp\left(-\int^t \bar{f}_n(s)\, ds \right) \right\},$$

$\int^t f$ indicating any indefinite integral of f. Simple calculation verifies that

$$X(t) = k(t, D^{-1}(t)Y(t)) \tag{3.7}$$

and

$$dY(t) = \tilde{G}(t, Y(t))\, dW(t) \tag{3.8}$$

where

$$\tilde{G}(t, Y(t)) = [Dh_x G](t, k(t, D^{-1}(t)Y(t))). \quad \square$$

Consider now the scalar equation

$$dX = f(X)\, dt + g(X)\, dW. \tag{3.9}$$

If $\tilde{g} \equiv 1$ is chosen a priori for simplicity, then (3.1) to (3.3) yield $Y(t) = W(t)$ and consequently

$$X(t) = \varphi(W(t), Z(t))$$

where

$$\frac{dZ(t)}{dt} = \tilde{f}(W(t), Z(t)).$$

Then, from (3.4) and (3.5), $\tilde{f}$ and φ must satisfy

$$\varphi_z \tilde{f} + \tfrac{1}{2}\varphi_{yy} = f(\varphi) \tag{3.10}$$

and

$$\varphi_y = g(\varphi). \tag{3.11}$$

Since from (3.11), $\varphi_{yy} = g'(\varphi)\varphi_y = g'(\varphi)g(\varphi)$, (3.10) can be written

$$\varphi_z \tilde{f} = g(\varphi)A(\varphi) \tag{3.12}$$

where

$$A(\varphi) = \frac{f(\varphi)}{g(\varphi)} - \tfrac{1}{2}g'(\varphi)$$

(as in Section 4.1). If $A(\varphi) = 0$ [a special case of the reducibility condition (1.1)], then φ can be taken as any solution of (3.11) independent of z and the simulation amounts to

$$X(t) = \varphi(W(t)).$$

In general, a necessary condition for a smooth solution φ of (3.11), (3.12) is the consistency requirement $\varphi_{yz} = \varphi_{zy}$, which is equivalent to the exactness condition

$$[g(\varphi)]_z = [g(\varphi)A(\varphi)/\tilde{f}]_y. \tag{3.13}$$

When $A \neq 0$, it is not difficult to see that the choice

$$\tilde{f}(y,z) = \exp \int^y [gA'/A](\varphi(u,z))\,du \tag{3.14}$$

achieves (3.13). Hence any solution φ of (3.11), (3.12) with $\tilde{f}$ given by (3.14) implements simulation of (3.9).

> *EXAMPLE 4.6*　For the stochastic equation
>
> $$dX = X^2\,dt + dW \tag{3.15}$$
>
> the choices $\varphi(y,z) = y + z$, $\tilde{f}(y,z) = (y+z)^2$ meet the conditions (3.11), (3.12), and (3.14). The solution of (3.15) can be written, therefore, as
>
> $$X(t) = W(t) + Z(t)$$
>
> where $Z(t)$ satisfies the Riccati-type equation
>
> $$\frac{dZ(t)}{dt} = [Z(t) + W(t)]^2.$$

Recent work (Doos, 1977; Krener, 1979; Krener and Lobry, 1981; Kunita, 1980; Lamperti, 1964; Sussman, 1978) indicates a relationship between various notions of solving stochastic differential equations

$$dX = f(X)\,dt + G(X)\,dW$$

and the structure of the Lie algebra generated by the vector fields $\{f, g^1,$
$\ldots, g^m\}$, where the g^j denote the columns of the matrix G; the product
operation in this algebra is defined by the Lie bracket

$$[h, k] = h\frac{dk}{dx} - k\frac{dh}{dx}.$$

For example, Krener and Lobry (1981) have shown that if the Lie algebra
generated by $\{g^1, \ldots, g^m\}$ is finite dimensional, then the stochastic system
can be simulated by a decoupled system.

EXERCISES

4.1 Show that the polynomial functions $f(x)$ of degree n for which the
stochastic differential equation

$$dX = f(X)\,dt + cX\,dW$$

is reducible have the form

$$f(x) = ax^n + bx$$

where a, b, and c are arbitrary constants.

4.2 Under what conditions on the parameters a, b, m, and n is the stochastic differential equation

$$dX = aX^n\,dt + bX^m\,dW$$

reducible?

4.3 Verify (1.29).

4.4 Write out the proof of Theorem 4.4.

4.5 Verify (2.4).

4.6 Check that $\tilde{f}$ given by (3.14) satisfies (3.13).

4.7 Verify that (3.15) is not reducible.

5

Qualitative Theory of Stochastic Differential Equations

5.0 INTRODUCTION

Generally, differential equations cannot be solved by quadratures; explicit expressions for solutions as functions of one or more independent variables may not exist. Even when obtainable, closed-form expressions for solutions may be sufficiently complicated as to prevent ascertaining fundamental solution properties. Yet the usefulness of differential equations in applications depends, to a large extent, on the feasibility of determining basic aspects of the nature of solutions for broad classes of equations. The qualitative theory of differential equations encompasses techniques and methods that permit investigating the general behavior of solutions directly from the form of the differential equations and any information available about initial or boundary data. Consequently, estimating the equation itself in some manner is usually involved. Quantitative methods, namely, the numerical approximation of solutions, on the other hand, give useful information for specific equations and choices of initial boundary data. Modern differential equations theory, much of which is qualitative theory, had its beginnings at the end of the last century with the work of Poincaré on celestial mechanics and Lyapunov's study of the stability of motions. The development of topological dynamics in this century by Birkhoff and others followed in the geometric vein of Poincaré's work and established a good deal of the foundation of qualitative theory. The differential inequality-comparison principle techniques, contributed to by many researchers and so prevalent in the qualitative theory, can be viewed, on the other hand, as extensions of the analytical methods of Lyapunov.

In this chapter, boundedness, stability, and uniqueness of solutions of stochastic differential equations are discussed through the Lyapunov-type approach. For ordinary differential equations in particular the corresponding results embody the simplest version of integrating a differential inequality involving an auxiliary function. That is, one deduces, from the nonpositivity of its derivative, that a Lyapunov-type function applied to a solution is a nonincreasing function. A key point here which makes this approach useful for qualitative results is that only the form of the Lyapunov function and the differential equation, and not the explicit form of the solution, are required to establish derivative nonpositivity. Properties of the solutions are determined then from those possessed by the Lyapunov-type function. In the stochastic case one uses Ito's formula, once again, to provide an expression for the stochastic differential of the Lyapunov function applied to a solution of the stochastic differential equation. From nonpositivity of the drift portion of this expression, one concludes that the Lyapunov function applied to a solution is a supermartingale. Analogous results for stochastic differential equations then follow from the properties of the Lyapunov function and the supermartingale theory.

5.1 BOUNDEDNESS OF SOLUTIONS

Let $f(t, x)$ be a continuous R^n-valued function defined for $t \in [t_0, T]$ and $x \in R^n$. A solution $x(t; t_0, x_0)$ of the deterministic initial value problem

$$\frac{dx}{dt} = f(t, x) \tag{1.1}$$

$$x(t_0) = x_0 \tag{1.2}$$

is *bounded* if there exists a constant $\beta = \beta(t_0, x_0) > 0$ such that

$$|x(t; t_0, x_0)| \leq \beta, \qquad \text{all } t \geq t_0. \tag{1.3}$$

Solutions of (1.1) are *uniformly bounded* if β is independent of $t_0 > 0$ and if for each $\alpha > 0$, there is a constant $\beta_\alpha > 0$ such that

$$|x(t; t_0, x_0)| \leq \beta_\alpha, \qquad t \geq t_0, \text{ all } t_0, \text{ and } |x_0| < \alpha. \tag{1.4}$$

The following theorem gives a condition in terms of Lyapunov-type functions for boundedness of solutions $x(t)$ of (1.1).

Throughout this chapter such functions V will be assumed to be nonnegative and continuously differentiable on an appropriate set. Key conditions in Lyapunov-type theorems are stated in terms of the derivative of V

along solution curves of (1.1),

$$\frac{dV(t, x(t))}{dt} = \dot{V}(t, x(t)) \equiv \left[\frac{\partial V}{\partial t} + \nabla_x V \cdot f \right] (t, x(t)), \tag{1.5}$$

which makes possible their verification without explicit knowledge of $x(t)$.

THEOREM 5.1   Suppose for $t > 0$ and $|x| \geq K$, some constant K, there exists a nonnegative continuously differentiable function $V(t, x)$ which satisfies the conditions

(i) $a(|x|) \leq V(t, x) \leq b(|x|)$, where a and b are continuous, and $a(r) \to \infty$ as $r \to \infty$.

(ii) $\dot{V}(t, x) \leq 0$.

Then solutions of (1.1) are uniformly bounded.

     The proof of this as well as the other results mentioned in this chapter for the ordinary initial value problem (1.1), (1.2) may be found in Yoshizawa (1966), for example.

     Throughout the next three sections, the following stochastic system, under conditions guaranteeing existence and uniqueness of solutions to initial value problems, is considered:

$$dX(t) = f(t, X(t)) \, dt + G(t, X(t)) \, dW(t), \tag{1.6}$$

where f is a continuous n-vector-valued function for $t \in [t_0, T]$, $x \in R^n$, and G is a continuous $n \times m$-matrix-valued function on the same set; f and G are at least locally Lipschitz in x; and $W = \{W_j\}$, is an m-dimenionsal Wiener process with independent component processes. A natural analogue of (1.3) for solutions of (1.6) is stochastic boundedness: the solution $X(t) = X(t; t_0, x_0)$ of (1.6) satisfying $X(t_0) = x_0$ is *stochastically bounded* (or bounded in probability) provided for each $\varepsilon > 0$, there exists a $\beta_\varepsilon = \beta_\varepsilon(t_0, x_0) > 0$ such that

$$\inf_{t_0 \leq t \leq T} P(|X(t)| \leq \beta_\varepsilon) > 1 - \varepsilon. \tag{1.7}$$

Uniform stochastic boundedness is defined analogously to (1.4): β_ε in (1.7) depends only on x_0. Stochastic boundedness implies that the process does not exhibit trends toward ever larger values and that the average frequency of fluctuations to large values does not increase with time, that is, essentially, that large values for X are observed infrequently. A sufficient condition for stochastic boundedness is that for some $\alpha > 0, E|X(t)|^\alpha$ is bounded on the interval $[t_0, T]$; this follows by Markov's inequality. A stronger form of boundedness for solutions of (1.6) (which might be called stochastic sample path boundedness) results from the following stochastic counterpart of Theorem 5.1. Notice that the only change in assumptions is that $\dot{V}(t,x)$ is replaced by the expression

$$LV(t,x) = \left[\frac{\partial V}{\partial t} + \nabla_x V \cdot f + \frac{1}{2}\operatorname{tr}(GG^T V_{xx})\right](t,x). \qquad (1.8)$$

THEOREM 5.2 Suppose there is a nonnegative continuous function $V(t,x)$ with continuous partial derivatives $\partial V/\partial t$, $\partial V/\partial x_i$, and $\partial^2 V/\partial x_i \partial x_j$ which satisfies the conditions

(i) $a(|x|) \leq V(t,x) \leq b(|x|)$, where a and b are continuous, and $a(r) \to \infty$ as $r \to \infty$

(ii) $LV(t,x) \leq 0$, for $t > 0$ and $|x| \geq K$, some constant K.

Let $\tau = \tau_K$ be the first exit time from $\{|x| > K\}$ for the solution $X(t)$ of (1.6) with $X(t_0) = x_0$, $|x_0| > K$. Then for each $\varepsilon > 0$, there exists a $\beta_\varepsilon = \beta_\varepsilon(x_0) > 0$ such that

$$P\left(\sup_{t_0 \leq t \leq \tau} |X(t)| \leq \beta_\varepsilon\right) > 1 - \varepsilon. \qquad (1.9)$$

Proof: Let τ_N denote the first exit time of the solution $X(t)$ from the annular region $K < |x| < N$, for any positive integer $N > K$. Set $\tau_N(t) = \tau_N \wedge t = \min(\tau_N, t)$, and let $X_N(t) = X(\tau_N(t))$ denote the corresponding stopped process. By Ito's formula, for $t_0 \leq s \leq t$,

$$V(t, X_N(t)) - V(s, X_N(s)) = \int_s^t LV(r, X_N(r))\, dr$$

$$+ \int_s^t [\nabla_x V G](r, X_N(r))\, dW(r)$$

$$\leq \int_s^t [\nabla_x V G](r, X_N(r))\, dW(r). \qquad (1.10)$$

by (ii). Let $\mathcal{A}(t)$ be the σ-algebra generated by the solution $\{X(s): s \leq t\}$; taking conditional expectation given $\mathcal{A}(s)$ in (1.10)

$$E(V(t, X_N(t)) - V(s, X_N(s)) \mid \mathcal{A}(s)) \leq 0,$$

that is, $V(t, X_N(t))$ is a supermartingale, each $N > K$. The positive supermartingale inequality [(3.6a) of Chapter 2] implies, for any $\alpha > 0$,

$$P\left(\sup_{[t_0, T]} V(t, X_N(t)) > \alpha\right) \leq \frac{1}{\alpha} V(t_0, x_0).$$

for any $T > t_0$. Letting $N \to \infty$, noting that $\tau_N \to \tau$ and that T is arbitrary, one has, then,

$$P\left(\sup_{[t_0,\tau]} V(t, X(t)) > \alpha\right) \leq \frac{1}{\alpha} V(t_0, x_0).$$

Making use of (i), it follows that

$$P\left(\sup_{[t_0,\tau]} a(X(t)) > \alpha\right) \leq \frac{1}{\alpha} b(|x_0|). \tag{1.11}$$

Since $a(r) \to \infty$ as $r \to \infty$, there exists $\beta = \beta_\alpha$ such that $|x| > \beta$ implies $a(x) > \alpha$, and so (1.11) yields

$$P\left(\sup_{[t_0,\tau]} |X(t)| > \beta\right) \leq \frac{1}{\alpha} b(|x_0|).$$

which verifies (1.9). □

 EXAMPLE 5.1   Consider the n-dimensional system

$$dX(t) = f(X(t))\, dt + G(X(t))\, dW(t). \tag{1.12}$$

The function

$$V(x) = \ln |x|,$$

$x^T = (x_1, \ldots, x_n)$, satisfies the conditions of Theorem 5.2 for $|x|^2 = \sum_{i=1}^n x_i^2 \leq 1$, except possibly (ii). Here

$$LV(x) = \frac{1}{|x|^2} Q(x) \tag{1.13}$$

where

$$Q(x) = x^T f(x) + \tfrac{1}{2} \operatorname{tr}(G(x)G(x)^T) - \frac{1}{|x|^2} x^T G(x) G(x)^T x.$$

So if

$$Q(x) \leq 0 \tag{1.14}$$

for $|x| \geq K$, some constant K, the conditions of Theorem 5.2 are met and so solutions of (1.12) satisfy (1.9).

For the scalar case of (1.12), with G replaced by g, condition (1.14) reduces to

$$2xf(x) \leq g^2(x), \qquad |x| \geq K. \tag{1.15}$$

The linear equation

$$dX = aX \, dt + bX \, dW \tag{1.16}$$

illustrates that (1.15) is best possible, since the solution

$$X(t) = X(0) \exp\left\{ \left(a - \frac{b^2}{2}\right) t + bW(t) \right\}$$

is unbounded as $t \to \infty$, w.p. 1, if $2a > b^2$.

If the hypotheses of Theorem 5.2 hold for all $x \in R^n$, the conclusion becomes

$$P(|X(t)| \leq \beta_\varepsilon, \text{ all } t \geq t_0) > 1 - \varepsilon. \tag{1.17}$$

In particular, explosions of the process $X(t)$ to ∞ (in finite or infinite time) are impossible. The *explosion time* τ_e of the process $X(t)$ is the unique random stopping time such that

(a) $X(t)I_{\{t < \tau_e\}}$ is nonanticipating.

(b) $X(t)$ solves (1.6) for $t < \tau_e$.

(c) $\displaystyle\lim_{t \to \tau_e} |X(t)| = +\infty.$

$$[\text{If } |X(t,\omega)| \not\to \infty, \text{ set } \tau_e(\omega) = +\infty.] \tag{1.18}$$

Equivalently, if $\{\tau_k\}$ is the sequence of first exit times of $X(t)$ from the balls of radius k about the origin, say, in R^n, then

$$\tau_e = \lim_{k \to \infty} \tau_k$$

The process $X(t)$ is said to *explode to* ∞ (in finite time) if

$$P(\tau_e < \infty) > 0. \tag{1.19}$$

On the other hand, when $P(\tau_e < \infty) = 0$, the process $X(t)$ is called *regular*. Regularity is the stochastic analogue of global existence for solutions of differential equations. It has been noted that the coefficient functions in (1.6) satisfying global Lipschitz conditions suffices to insure global existence.

Regularity can also be established under the somewhat weaker hypothesis on V,

$$LV \leq cV \tag{1.20}$$

for some positive constant c; that is, if the hypotheses of Theorem 5.2 hold for all $t > 0$ and all $x \in R^n$, except that (1.20) replaces (ii) (Has'minskii [47], Chapter 3, Theorem 4.1, pp. 84–85). Indeed, in this case, the function

$$\tilde{V}(t,x) = V(t,x) \exp\{-c(t - t_0)\} \tag{1.21}$$

satisfies

$$L\tilde{V} \leq 0,$$

and, as in the proof of Theorem 5.2, $\tilde{V}(t, X_k(t))$ is a supermartingale for each k, that is,

$$E\tilde{V}(t, X_k(t)) \leq \tilde{V}(t_0, x_0) = V(t_0, x_0).$$

From (1.21), then one has

$$EV(t, X_k(t)) \leq \exp\{c(t - t_0)\} V(t_0, x_0). \tag{1.22}$$

Since

$$EV(t, X_k(t)) \geq \left[\min_{u \geq t_0, |x| \geq k} V(u, x)\right] P(\tau_k \leq t), \tag{1.23}$$

(1.22) leads to the estimate

$$P(\tau_k \leq t) \leq \frac{e^{c(t-t_0)} V(t_0, x_0)}{\min_{u \geq t_0, |x| \geq k} V(u, x)}. \tag{1.24}$$

Radial unboundedness of V, then, implies that X is stochastically bounded on $[t_0, t]$, and further, by letting $k \to \infty$ in (1.24), that

$$P(\tau_e \leq t) = 0. \tag{1.25}$$

Since t is arbitrary, (1.25) establishes regularity. In this discussion R^n may be replaced by any open set $D \subseteq R^n$ and τ_e by the first exit time τ_D from D. The subsequent generalization yields a sufficient condition for D-invariance for solutions of (1.6), summarized in the following theorem.

THEOREM 5.3   Let D and D_n, each postive integer n, be open sets in R^n with

$$D_n \subseteq D_{n+1}, \quad \overline{D}_n \subseteq D, \quad \text{and } D = \bigcup_n D_n,$$

and suppose f and G satisfy the existence and uniqueness conditions for solutions of (1.6) on each set $t > 0$, $x \in D_n$.

Suppose, further there is a nonnegative continuous function $V(t, x)$ with continuous partial derivatives $\partial V/\partial t$, $\partial V/\partial x_i$, and $\partial^2 V/\partial x_i \partial x_j$ and satisfying (1.20) for $t > 0$, $x \in D$. If also

$$\inf_{t>0, x \in D \setminus D_n} V(t, x) \longrightarrow \infty \qquad \text{as } n \longrightarrow \infty$$

then, for any X_0 independent of $W(t)$ such that

$$P(X_0 \in D) = 1,$$

there is a unique solution $X(t)$ of (1.6) with $X(0) = X_0$, and $X(t) \in D$ all $t > 0$, that is, $P(\tau_D = \infty) = 1$.

Examples are given in sections 5.4, 6.2, and 6.3.

5.2 STABILITY

Stability for an equilibrium or other designated solution of a differential equation constitutes an important qualitative concept for applications. It asserts, roughly speaking, that solutions which start near the designated one remain close to that particular solution for all time. More precisely, consider the differential equation (1.1) and assume also that $f(t, 0) = 0$, all t; then $x(t; t_0, 0) \equiv 0$ is a solution of (1.1). This solution is said to be *stable* if given $\varepsilon > 0$, there is a $\delta = \delta(s, \varepsilon) > 0$ such that for all $t > s \geq t_0$

$$|x(t; s, x)| \leq \varepsilon \tag{2.1}$$

whenever $|x| < \delta$. If, in addition, there exists a $\bar{\delta} = \bar{\delta}(s) > 0$ such that

$$x(t; s, x) \longrightarrow 0 \qquad \text{as } t \longrightarrow \infty \tag{2.2}$$

whenever $|x| < \bar{\delta}$, then the zero solution is *asymptotically stable*. If δ (resp. $\bar{\delta}$) is not dependent on s, stability (resp. asymptotic stability) is said to be *uniform*. If (2.2) holds for all x_0, then $x(t) \equiv 0$ is referred to as *globally asymptotically stable* or *asymptotically stable in the large*. The basic Lyapunov stability theorems, analogous to Theorem 5.1, follow the next definition.

Let U be a neighborhood of the origin in R^m, and let $h : U \to R_+ = [0, \infty)$; if $h(0) = 0$ and $h(x) > 0$, $x \in U$, $x \neq 0$, then h is said to be *positive definite*; h is called *negative definite* if $-h$ is positive definite.

THEOREM 5.4   Assume, for $t > 0$ and $|x| \leq K$, some constant K, there exists a continuously differentiable function $V(t, x)$ which satisfies the conditions

(i) $a(|x|) \leq V(t, x) \leq b(|x|)$ where the functions a and b are continuous and positive definite on R_+,

(ii) $\dot{V}(t, x) \leq 0$.

Then the zero solution of (1.1) is (uniformly) stable.

THEOREM 5.5   If the assumptions in Theorem 5.4 hold, except that (ii) is replaced by

(ii′) $\dot{V}(t,x) \leq -c(|x|)$, where c is a continuous, positive definite function on R_+,

then the zero solution of (1.1) is (uniformly) asymptotically stable. Furthermore, if (i) and (ii′) hold for all x, and the function a appearing in (i) also satisfies $a(r) \to \infty$ as $r \to \infty$ (V is called *radially unbounded*), then $x(t) \equiv 0$ is globally asymptotically stable.

Theorems 5.4 and 5.5 are well known and can be found in many places in the literature. The smoothness assumptions, in particular differentiability at $x = 0$, on V in Theorem 5.4 and 5.5 can be relaxed somewhat.

EXAMPLE 5.2   Consider the system

$$x'(t) = Ax(t) + h(t, x(t)) \tag{2.3}$$

where A is an $n \times n$ matrix and the n-vector function h is continuous and satisfies

$$\lim_{|x| \to 0} |h(t,x)|/|x| = 0, \qquad \text{uniformly in } t. \tag{2.4}$$

Theorem 5.4 yields that the zero solution of (2.3) is asymptotically stable if all eigenvalues of A have negative real part. Indeed here, for a given positive definite matrix C (the quadratic form, $x^T C x$ is a positive definite function on R^n), a unique symmetric positive definite solution B of the matrix equation

$$A^T B + BA = -C$$

exists. It is easy to check that the quadratic function $V(x) = x^T B x$ satisfies the conditions in Theorems 5.4 and 5.5.

Consider now, once again, the stochastic equation (1.6). Throughout this and the next section it will be assumed that f and G in (1.6) satisfy, in addition to the existence and uniqueness conditions given with (1.6) in the previous section, that

$$f(t,0) = 0 \qquad \text{and} \qquad G(t,0) = 0 \qquad \text{all } t \geq t_0;$$

the latter ensures that the solution of (1.6) satisfying $X(t_0) = 0$ is the identically zero solution $X(t) = 0$, all $t \geq t_0$. Among the various stability notions proposed in this setting, the concept stochastic stability, developed by Has'minskii, has received the widest acceptance.

The solution $X(t) \equiv 0$ is said to be *stochastically stable* if for every $\varepsilon > 0$ and $s \geq t_0$

$$\lim_{x \to 0} P \left(\sup_{[s,\infty)} |X(t; s, x)| \geq \varepsilon \right) = 0, \tag{2.5}$$

where $X(t; s, x)$ denotes the solution of (1.6) satisfying the constant initial condition $X(s) = x$; the stability is said to be uniform if the limit in (2.5) is uniform in s. $X(t) \equiv 0$ is *stochastically asymptotically stable* if, in addition,

$$\lim_{x \to 0} P \left(\lim_{t \to \infty} X(t; s, x) = 0 \right) = 1, \qquad s \geq t_0 \tag{2.6}$$

and *stochastically asymptotically stable in the large* if, further,

$$P \left(\lim_{t \to \infty} X(t; s, x) = 0 \right) = 1, \qquad \text{all } x \in R^n. \tag{2.7}$$

Has'minskii uses the terminology "stable in probability" for (2.5), which is justified since (2.5) is equivalent to the condition: For each $s \geq t_0$ and ε_1, $\varepsilon_2 > 0$ there exists $\delta > 0$ such that

$$P \left(\sup_{t \geq s} |X(t; s, x)| \geq \varepsilon_1 \right) \leq \varepsilon_2 \tag{2.8}$$

whenever $|x| < \delta$.

For stability, just as illustrated for boundedness in the previous section, stochastic Lyapunov theorems can be obtained from corresponding deterministic results by replacing $\dot{V}$ by the expression LV given by (1.8).

THEOREM 5.6   Assume for $t > 0$ and some constant K

(i)   There is a continuous function $V(t, x)$ satisfying, for $|x| \leq K$,

$$a(|x|) \leq V(t, x) \leq b(|x|),$$

where the functions a and b are continuous and positive definite on R_+.

(ii)   The partial derivatives $\partial V / \partial t$, $\partial V / \partial x_i$, and $\partial^2 V / \partial x_i \partial x_j$ exist, are continuous, and

$$LV(t, x) \leq 0 \tag{2.9}$$

for $0 < |x| \leq K$, where L is given by (1.8).

Then the zero solution of (1.6) is (uniformly) stochastically stable.

THEOREM 5.7   Suppose the assumptions in Theorem 5.6 hold, except that (2.9) is replaced by

$$LV(t, x) \leq -c(|x|) \tag{2.10}$$

where the function c is continuous and positive definite on R_+. Then the zero solution of (1.6) is the stochastically asymptotically stable. Furthermore, if these assumptions are satisfied for all x ($\neq 0$) and the function a appearing in (i) also satisfies

$$a(r) \longrightarrow \infty \qquad \text{as } r \longrightarrow \infty,$$

then $X(t) \equiv 0$ is stochastically asymptotically stable in the large.

Proof of Theorem 5.6:   For any $r < K$, set

$$V_r = \inf_{r \leq |x| \leq K} V(t, x);$$

$V_r > 0$. For $0 < |x| \leq r$, consider the solution $X(t) = X(t; s, x)$, satisfying $X(s) = x$, let τ_r be its first exit time from the ball of radius r about the origin, and let

$$\tau_r(t) = \tau_r \wedge t = \min\{\tau_r, t\}.$$

The main part of the proof establishes that

$$EV(\tau_r(t), X(\tau_r(t))) \leq V(s, x) \tag{2.11}$$

for $t > s$ and $0 < |x| < r$. As in the proof of Theorem 5.2, from (2.11) it follows that

$$P\left(\sup_{s \leq u} |X(u)| > r\right) \leq \frac{1}{V_r} V(s, x). \tag{2.12}$$

Taking into account that V is continuous and $V(s, 0) = 0$, (2.12) implies the conclusion of Theorem 5.6. Therefore, it remains to show (2.11).

The assumption that V is not smooth on $x = 0$ causes some slight technical difficulties in that Ito's formula cannot be immediately applied to V on the interval $[s, \tau_r(t)]$ to obtain the desired result in the same manner as the Theorem 5.2. The problems are overcome by showing that

(i) τ_δ, in place of τ_r, satisfies (2.11) for each $0 < \delta < r$, where $\delta < |x| < r$ and

$$\tau_\delta(t) = \min\{t, \text{ first exit time of } X(t; s, x) \text{ from } B_r \setminus B_\delta\}.$$

(ii) $P(\tau_\delta(t) \to \tau_r(t), \text{ as } \delta \to 0) = 1.$

Since $X \neq 0$ on the interval $[s, \tau_\delta(t)]$, and V is bounded, Ito's formula can be applied to $V(\cdot, X(\cdot))$ on $[s, \tau_\delta(t)]$, (2.9) used, and expectation taken to arrive at

$$EV(\tau_\delta(t), X(\tau_\delta(t))) \leq V(s, x); \tag{2.13}$$

that is, the method employed in the proof of Theorem 5.2 can be applied here to verify (i).

Toward obtaining (ii), let $Y(t) = |X(t)|^\beta$, for a real number β; as in the proof of Theorem 3.8, it can be verified that

$$\mathrm{E}Y(\tau_\delta(t)) \leq |x|^\beta + kE \int_s^{\tau_\delta(t)} Y(u)\, du \tag{2.14}$$

where k is a constant depending on β and the Lipschitz constant for the functions f and G. Since $u = \tau_\delta(u)$ for $u < \tau_\delta(t)$, (2.14) can be written

$$EY(\tau_\delta(t)) \leq |x|^\beta + kE \int_s^{\tau_\delta(t)} Y(\tau_\delta(u))\, du. \tag{2.15}$$

The Bellman-Gronwall inequality (Lemma 3.2) indicates that

$$E|X(\tau_\delta(t))|^\beta \leq |x|^\beta \exp\{k(t - s)\} \tag{2.16}$$

follows from (2.15). Now $\tau_\delta(t) \leq \tau_r(t)$ w.p. 1, and

$$P(\tau_\delta(t) < \tau_r(t)) = P(|x(\tau_\delta(t))|^{-1} = \delta^{-1}) \leq \delta E|X(\tau_\delta(t))|^{-1}.$$

So, using (2.16) with $\beta = -1$,

$$P(\tau_\delta(t) < \tau_r(t)) \leq \frac{\delta}{|x|} \exp\{k(t - s)\}. \tag{2.17}$$

Since $\tau_\delta(t) \leq \tau_r(t)$ w.p. 1, and since $\tau_{\delta_1}(t) = \tau_r(t)$ implies $\tau_\delta(t) = \tau_r(t)$ for all $\delta < \delta_1$, (ii) follows from (2.17).

Finally letting $\delta \to 0$ in (2.13) obtains (2.11) which completes the proof. $\square$

Proof of Theorem 5.7: Retain the notation in the previous proof. $V(\tau_r(t), X(\tau_r(t)))$ is a supermartingale. The supermartingale convergence theorem [(3.8) of Chapter 2] implies that the w.p. 1

$$\lim_{t \to \infty} V(\tau_r(t), X(\tau_r(t))) \tag{2.18}$$

exists. Let $\Omega_x = \{\omega \in \Omega : \tau_r = \infty\}$. Stochastic stability of the solution $X(t) \equiv 0$ implies that

$$P(\Omega_x) \longrightarrow 1 \qquad \text{as } x \longrightarrow 0. \tag{2.19}$$

Now, once again, letting τ_δ be the first exit time of the process $X(t; s, x)$ from $B_r \setminus B_\delta$, where

$$\delta < |x| < r,$$

except for a subset of probability zero, the paths in Ω_x satisfy

(a) $\tau_\delta < \infty$.

(b) $|X(\tau_\delta)| = \delta$.
$$\text{(2.20)}$$

Together these imply that

$$\liminf_{t\to\infty} |X(t; s, x)| = 0 \tag{2.21}$$

for $\omega \in \Omega_x$, except for a set of probability zero.   Since on Ω_x, $\tau_r(t)$ $(= \tau_r \wedge t) = t$, so one can write (2.21) as

$$\liminf_{t\to\infty} |X(\tau_r(t))| = 0. \tag{2.22}$$

Assumption (i) $(V \le b(|x|)$ part) implies

$$\liminf_{t\to\infty} V(\tau_r, X(\tau_r(t))) = 0. \tag{2.23}$$

Then (2.18) and (2.23) together imply that, for all sample paths $\omega \in \Omega_x$, except for a subset of probability zero,

$$V(t, X(t; s, x)) = V(\tau_r(t), X(\tau_r(t))) \longrightarrow 0 \qquad \text{as } t \longrightarrow \infty;$$

and using (i) $(a(|x|) \le V$ part), one has

$$X(t) \longrightarrow 0 \qquad \text{as } t \longrightarrow \infty. \tag{2.24}$$

Noting (2.19), (2.24) establishes the result. So to complete the proof, it remains to verify (2.20), in particular (2.20a); (2.10) is used here. Apply Ito's formula to the process $V(\tau_\delta(t), X(\tau_\delta(t)))$ and take expected value to obtain

$$EV(\tau_\delta(t), X(\tau_\delta(t))) = V(s, x) + E \int_s^{\tau_\delta(t)} LV(\tau_\delta(u), X(\tau_\delta(u)))\, du.$$
$$\text{(2.25)}$$

Since $X(\tau_\delta(t)) \in \overline{B}_r \setminus B_\delta$, all $t \ge s$, the left side of (2.25) cannot be less than V, and (2.10) implies that, for all $u \ge s$,

$$LV(\tau_\delta(u), X(\tau_\delta(u))) \le -c(|X(\tau_\delta(u))|) \le -c(\delta);$$

so (2.25) implies

$$V_\delta \le V(s, x) + E \int_s^{\tau_\delta(t)} -c(\delta)\, du = V(s, x) - c(\delta)[E(\tau_\delta(t)) - s]$$

or

$$E(\tau_\delta(t)) \le \frac{V(s, x) - V_\delta}{c(\delta)} + s. \tag{2.26}$$

Since (2.26) holds for all $t \geq s$, for sufficiently large t,

$$\tau_\delta(t)(= \tau_\delta \wedge t) = \tau_\delta$$

and so (2.26) establishes (2.20a). Since only paths in Ω_x are considered, (2.20b) follows immediately, and the the proof of the first statement in Theorem 5.7 is complete. The proof of the rest of the theorem follows similarly and is left to the reader (Exercise 5.5). □

EXAMPLE 5.3 Consider the scalar equation

$$dX(t) = f(t, X(t))\, dt + g(t, X(t))\, dW(t) \tag{2.27}$$

where, in addition to the existence and uniqueness conditions, the functions f and g satisfy, for some constants a and b,

$$\begin{aligned} f(t,x) &= ax + \bar{f}(t,x) \\ g(t,x) &= bx + \bar{g}(t,x) \end{aligned} \tag{2.28}$$

with

$$\lim_{|x| \to 0} \frac{|\bar{f}(t,x)| + |\bar{g}(t,x)|}{|x|} = 0, \qquad \text{uniformly in } t.$$

Then the $X(t) \equiv 0$ solution of (2.27) is stochastically asymptotically stable if $a - \frac{1}{2}b^2 < 0$. Indeed, to see this, consider the function $V(x) = |x|^k$, for $k > 0$. For this choice of V, it follows that

$$\begin{aligned} LV(x) &= \left\{ a + \frac{\bar{f}(t,x)}{x} + \frac{1}{2}(k-1)\left[b + \frac{\bar{g}(t,x)}{x} \right]^2 \right\} k|x|^k \\ &= \left\{ a - \frac{1}{2}b^2 + \frac{\bar{f}(t,x)}{x} + \frac{1}{2}kb^2 \right. \\ &\quad \left. + (k-1)\left[\frac{b\bar{g}(t,x)}{x} + \frac{\bar{g}^2(t,x)}{2x^2} \right] \right\} k|x|^k. \end{aligned}$$

Since k and r can be chosen so small by (2.28) such that

$$\left| \frac{\bar{f}(t,x)}{x} \right| + \frac{1}{2}kb^2 + \left| (k-1)\left[\frac{b\bar{g}(t,x)}{x} + \frac{\bar{g}^2(t,x)}{2x^2} \right] \right| < |a - \tfrac{1}{2}b^2|$$

on $B_r \setminus \{0\}$, there is a positive constant c such that

$$LV(x) \leq -cV(x) \tag{2.29}$$

on $B_r \setminus \{0\}$. The assertion, then, follows from Theorem 5.7.

For the corresponding linear equation,

$$dX(t) = aX(t)\,dt + bX(t)\,dW(t) \qquad (2.30)$$

the preceding analysis results in (2.29) holding for all $x \neq 0$. Since $V(x) = |x|^k$ is radially unbounded, it can be concluded that $X(t) \equiv 0$ is stochastically asymptotically stable in the large if $a - \frac{1}{2}b^2 < 0$, by the second part of Theorem 5.7. Recalling that the solution of (2.30) is

$$X(t; s, x) = x \exp\left\{ \left(a - \tfrac{1}{2}b^2\right)(t - s) + b(W(t) - W(s)) \right\},$$

and noting that $[W(t) - W(s)]/(t - s) \to 0$ as $t \to \infty$, it is clear that this result is sharp. That is, $X(t) \equiv 0$ is stochastically asymptotically stable in the large if $a - \frac{1}{2}b^2 < 0$ and unstable if $a - \frac{1}{2}b^2 \geq 0$.

The corresponding linearization or stability in the first approximation result for systems of stochastic differential equations is given in the next section.

5.3   FURTHER REMARKS ON STOCHASTIC STABILITY

Because of its importance in applications in engineering control problems, a rich mathematical theory of stochastic stability has developed over the past thirty years. In the previous section the basic Lyapunov function type theorems which provide sufficient conditions for stochastic stability were given. In this section some other aspects of this theory are briefly described. Detailed discussion of most of the remarks made here can be found in Has'minskii (1980).

First of all, note that in the proof of Theorem 5.7, negative definiteness of LV (2.10) was used only to guarantee finite time exit from the annular region $\delta < |x| < r$, for arbitrary small δ. Recall (see proof of Theorem 3.14) that the latter property also follows from the noise coefficient nondegeneracy condition:

$$\sum_{i,j=1}^{n} \left(\sum_{k=1}^{m} g_{ik}g_{jk}(t, x) \right) \xi_i \xi_j > M(x)|\xi|^2 \qquad \text{for each}$$

$\xi = \{\xi_i\} \in R^n$, $t \geq 0$, and all x in a neighborhood of 0,
where M is a continuous and positive definite
function in that neighborhood. $\qquad (3.1)$

So stochastic asymptotic stability holds if (2.10) is replaced by (2.9) and (3.1).

Converse theorems, of the type Malkin has given for ordinary differential equations, are also known for the stochastic case. For example, for autonomous systems

$$dX(t) = f(X(t))\, dt + G(X(t))\, dW(t), \tag{3.2}$$

stochastic stability of the trivial solution $X(t) \equiv 0$ together with the non-degeneracy condition (3.1) imply the existence of a Lyapunov function V satisfying the conditions of Theorem 5.5. Taking into account the previous remark, for autonomous systems (3.2) satisfying (3.1), then stochastic stability and stochastic asymptotic stability are equivalent.

Example 5.3 can be extended to the n-dimensional case via two results, namely, a linearization theorem and a criterion for stochastic stability in linear systems with constant coefficients. Denote by $g_k(t, x)$ the columns of the matrix $G(t, x)$. Then (1.6) can be written

$$dX(t) = f(t, X(t))\, dt + \sum_{k=1}^{m} g_k(t, X(t))\, dW_k(t). \tag{3.3}$$

Suppose there exist constant matrices A and B_k, $k = 1, \dots, m$, such that

$$\lim_{|x| \to 0} \frac{|f(t, x) - Ax| + \sum_{k=1}^{m} |g_k(t, x) - B_k x|}{|x|} = 0. \tag{3.4}$$

Consider the linear constant coefficient system

$$dX(t) = AX(t)\, dt + \sum_{k=1}^{m} B_k X(t)\, dW_k(t). \tag{3.5}$$

The linearization result states that the trivial solution $X(t) \equiv 0$ of (3.3) is stochastically asymptotically stable if the trivial solution of (3.5) is stochastically asymptotically stable. Now for the linear system (3.5) one can give a stability condition in terms of the expression

$$Q(y) = y^T A y + \tfrac{1}{2} \operatorname{tr} B(y) B(y)^T - \frac{1}{|y|^2} y^T B(y) B(y)^T y \tag{3.6}$$

where $B(y)$ is the $n \times m$ matrix with kth column $B_k y$. Assume that the nondegeneracy condition (3.1) holds for $B(y)$ in place of G. Let $X(t) = X(t; x)$, $x \neq 0$, be a solution of (3.5) and set $Y(t) = X(t)/|X(t)|$. The nondegeneracy condition implies that, for the process $Y(t)$ defined on the sphere $|y| = 1$, there exists a number α such that

$$\alpha = \lim_{t \to \infty} \int_0^t Q(Y(s))\, ds \qquad \text{w.p. 1} \tag{3.7}$$

from the strong law of large numbers. The relevance of (3.7) to the stability question can be seen if one applies Ito's formula to the process

$$Z(t) = \ln |X(t)|$$

and divides by t to obtain

$$\frac{Z(t) - Z(0)}{t} = \frac{1}{t} \int_0^t Q(Y(s)) \, ds + \frac{1}{t} \sum_{k=1}^m \int_0^t Y(s)^T B_k Y(s) \, dW_k(s).$$

$$(3.8)$$

Since the corresponding integrands are bounded,

$$\lim_{t \to \infty} \frac{1}{t} \int_0^t Y^T B_k Y \, dW_k = 0 \qquad \text{w.p. 1,}$$

for each k, similarly to (9.3) in Chapter 1.

So letting $t \to \infty$ in (3.8) gives

$$\lim_{t \to \infty} \frac{\ln |X(t)|}{t} = \alpha \qquad \text{w.p. 1.}$$

Therefore, if $\alpha < 0$,

$$P\left(\lim_{t \to \infty} |X(t)| = 0\right) = 1,$$

$$(3.9)$$

whereas if $\alpha > 0$,

$$P\left(\lim_{t \to \infty} |X(t)| = \infty\right) = 1.$$

Note that the condition

$$Q(y) < 0, \qquad \text{all } y \neq 0,$$

$$(3.10)$$

certainly implies $\alpha < 0$ and avoids computing it. Summarizing, for linear systems (3.5) under (3.1), the concepts stochastic stability, asymptotic stochastic stability, and asymptotic stochastic stability in the large, for the trivial solution $X(t) \equiv 0$, are all equivalent and implied by $\alpha < 0$. The stability condition (3.10) can be viewed as an extension of the inequality

$$a - \tfrac{1}{2} b^2 < 0$$

$$(3.11)$$

for (2.27) with (2.28), and (2.30), the scalar case of (3.3) to (3.5). Note that (3.11) does not require negative a and, more generally, (3.10) may hold even if A is not negative definite. This suggests the possibility of "stabilizing" an unstable deterministic system by introducing noise of sufficiently large intensity, provided the Ito interpretation of (3.3) and (3.5) can be justified.

Instability for stochastic differential equations, as well as stability, can be ascertained via Lyapunov functions, as illustrated in the next theorem.

THEOREM 5.8   Assume (3.1), (ii) of Theorem 5.6, and, in place of (i), let

$$a(|x|) \leq V(t, x)$$

hold for $t > 0$ and $0 < |x| \leq K$, where a is a continuous function on R_+ and $a(r) \to +\infty$ as $r \to 0$; then $X(t) \equiv 0$ is not stochastically stable, and, in fact

$$\left\{ \sup_{t>0} |X(t; s, x)| < K \right\}$$

has probability zero for all $s > 0$ and $0 < |x| < K$.

The proof is not difficult and is left to the reader (Exercise 5.7).

EXAMPLE 5.4   Consider, once again, the n-dimensional system

$$dX(t) = f(X(t))\, dt + G(X(t))\, dW(t). \tag{3.12}$$

The function

$$V(x) = -\ln |x|$$

satisfies the instability conditions in Theorem 5.8 if

$$LV(x) = -\frac{1}{|x|^2} Q(x) \leq 0,$$

$0 < |x| < K$, some $K > 0$, where, as in (1.13) and (3.6),

$$Q(x) = x^T f + \tfrac{1}{2} \operatorname{tr} GG^T - \frac{1}{|x|^2} x^T GG^T x.$$

So instability occurs if

$$Q(x) \geq 0 \tag{3.13}$$

sufficiently close to the origin. The scalar and scalar linear versions of (3.13) reduce to

$$2xf(x) \geq g^2(x), \qquad 0 < |x| \leq K, \text{ some } K > 0, \tag{3.14}$$

and

$$2a \geq b^2, \tag{3.15}$$

respectively.

Stability results discussed in this and the previous section are predicated, of course, on the existence of an equilibrium $\bar{x}$ for (1.6); this, requires, for all $t > 0$,

$$f(t, \bar{x}) = 0 \tag{3.16}$$

and

$$G(t, \bar{x}) = 0. \tag{3.17}$$

Without loss of generality $\bar{x}$ is taken to be zero, and then the pertinent question is whether or not the trivial solution $X(t) \equiv 0$ of (1.6) is stable. Notice that such a constant solution can be considered as a degenerate invariant distribution for (1.6) with Dirac delta density $\delta(x - \bar{x})$. Even when either (3.16) or (3.17) fails, the question of whether a (nontrivial) invariant distribution for (1.6) remains. Such a distribution serves as a stochastic analogue of a deterministic equilibrium, in this case. The density $P(x)$ of such a distribution, if it is sufficiently smooth, is an equilibrium solution of the forward (or Fokker-Planck) equation, that is,

$$0 = \sum_{i=1}^{n} \frac{\partial}{\partial x_i} [f_i(t, x) p(x)]$$

$$- \frac{1}{2} \sum_{i,j=1}^{n} \frac{\partial^2}{\partial x_i \partial x_j} \left[\sum_{k=1}^{m} g_{ik}(t, x) g_{jk}(t, x) p(x) \right] \qquad \text{for } x \in D, \quad (3.18)$$

where D is an open set in R^n containing the support of $p(x)$. Also, being a probability density, p must satisfy

$$p(x) \geq 0, \qquad x \in D \tag{3.19}$$

and

$$\int_D p(x) \, dx = 1. \tag{3.20}$$

In general, determining a solution of (3.18) to (3.20) is not easy. (In Sections 5.4 and 6.2 the scalar autonomous case is discussed and examples are given.) For autonomous equations (3.2) the following theorem gives, in terms of Lyapunov functions, a criterion for the existence of an invariant distribution, which is stable in a certain sense. Among its hypotheses is the condition that D is an invariant set for solutions of (3.2); hence stochastic boundedness (if $D = R^n$) or more generally Theorem 5.3 might be invoked to establish this requisite. Specifically, Theorem 5.9 is established by combining some results in Has'minskii (1980), namely: remark 2 following Theorem 4.1 of Chapter 3 (p. 86), Theorem 7.1 of Chapter 3 (p. 98), and generalization 4 of Theorem 7.1 in Chapter 4 (p. 134).

THEOREM 5.9   Suppose the conditions of Theorem 5.3 together with the following: For some positive integer n_0, there are positive constants M and c such that

(i) $\displaystyle\sum_{i,j=1}^{n} \left(\sum_{k=1}^{m} g_{ik}(x) g_{jk}(x) \right) \xi_i \xi_j \geq M |\xi|^2$ all $x \in \overline{D}_{n_0}$, $\xi \in R^n$.

(ii) $LV(x) \leq -c$, all $x \in D \setminus \overline{D}_{n_0}$.

Then there exists an invariant distribution $\tilde{P}$ with nowhere zero density in D such that for any Borel set $B \subseteq D$

$$P(t, x, B) \longrightarrow \tilde{P}(B) \qquad \text{as } t \longrightarrow \infty$$

where $P(t, x, B)$ is the transition probability $P(X(t) \in B | X(0) = x)$ for the solution X of the stochastic differential equation (3.2).

5.4 QUALITATIVE THEORY OF SCALAR DIFFUSION PROCESSES: BOUNDARY CLASSIFICATION

The qualitative theory of stochastic differential equations presented in this chapter can be considered as an extension of the boundary classification scheme for scalar homogeneous diffusion processes which was originated by Feller (1954). [See also Prohorov and Rozanov (1969) and Gihman and Skorohod (1972).] In this section a brief account of this scheme is given and its connection with stochastic boundedness and stability is pointed out.

The scalar homogenous diffusion process $X(t) = X(t; 0, x)$ taking values in the interval (x_1, x_2) and having drift coefficient f and diffusion coefficient g^2 solves the Ito-interpreted autonomous stochastic differential equation

$$dX = f(X) \, dt + g(X) \, dW \tag{4.1}$$

with initial value $X(0) = x \in (x_1, x_2)$. In (4.1), for simplicity, it is asumed that f and g are at least continuously differentiable on (x_1, x_2) [so that existence and uniqueness for initial value problems involving (4.1) holds], and that

$$g \neq 0, \qquad \text{on } (x_1, x_2). \tag{4.2}$$

Recall (see Section 3.5) that (4.2) guarantees that the first exit time $\tau_x[\alpha, \beta]$ of X from the interval $[\alpha, \beta] \subseteq (x_1, x_2)$, $x \in [\alpha, \beta]$, is finite with probability 1. Let τ_x be the (not necessarily finite) exit time from (x_1, x_2):

$$\tau_x = \lim_{n \to \infty} \tau_x[\alpha_n, \beta_n] \tag{4.3}$$

where

$$[\alpha_n, \beta_n] \subseteq [\alpha_{n+1}, \beta_{n+1}] \qquad \text{and} \qquad \bigcup_n [\alpha_n, \beta_n] = (x_1, x_2).$$

The boundary point x_i is called *attainable* (or *accessible*) if

$$P\left(\lim_{t \to \tau_x} X(t) = x_i, \tau_x < \infty\right) > 0, \tag{4.4}$$

and *unattainable* (or *inaccessible*) otherwise. More generally, if

$$P\left(\lim_{t \to \tau_x} X(t) = x_i\right) > 0, \tag{4.5}$$

x_i is called *attracting*, and *repelling* otherwise. Repelling boundaries are also called *natural* boundaries. Attainable boundaries must be attracting, and natural boundaries must be unattainable, but not conversely. Observe that the boundary point x_2, say, is a natural boundary if and only if, for $x \in [\alpha, \beta] \subseteq (x_1, x_2)$,

$$P(X(\tau_x[\alpha, \beta]) = \beta)$$

becomes arbitrarily small as $\beta \to x_2$.

The boundary classification can be characterized in terms of integrability conditions on expressions involving

$$\varphi(x) = \exp\left\{-\int_{x_0}^{x} \frac{2f(y)}{g^2(y)}\, dy\right\}, \tag{4.7}$$

$x_0 \in (x_1, x_2)$ arbitrary, over intervals near the boundary. In particular, the boundary point x_i is repelling if φ is not integrable over a neighborhood of x_i in (x_1, x_2), and attracting if φ is integrable over such a neighborhood. Indeed, setting

$$\psi(x) = \int_{x_0'}^{x} \varphi(y)\, dy \tag{4.8}$$

for arbitrary $x_0' \in (x_1, x_2)$,

$$P(X(\tau_x[\alpha, \beta] = \beta) = \frac{\psi(x) - \psi(\alpha)}{\psi(\beta) - \psi(\alpha)} = \frac{\int_\alpha^x \varphi(y)\, dy}{\int_\alpha^\beta \varphi(y)\, dy}, \tag{4.9}$$

which becomes arbitrarily small as $\beta \to x_2$ if

$$\psi(\beta) \longrightarrow \infty \qquad \text{as } \beta \longrightarrow x_2.$$

If, on the other hand, φ is integrable near x_2, that is,

$$\psi(\beta) \longrightarrow \psi(x_2)\ (< \infty) \qquad \text{as } \beta \longrightarrow x_2,$$

then

$$P\left(\lim_{t \to \tau_x[\alpha, \infty)} X(t) = x_2\right) = \lim_{\beta \to x_2} P(X(\tau_x[\alpha, \beta]) = \beta) = \frac{\int_\alpha^x \varphi(y)\, dy}{\int_\alpha^{x_2} \varphi(y)\, dy}.$$

(4.10)

The boundary point x_i is attainable if and only if

$$\xi(x) = \varphi(x) \int_{x_0'}^x [g^2(y)\varphi(y)]^{-1}\, dy$$

(4.11)

is integrable in a neighborhood of x_i. These results confirm the relationships between attainable and attracting, natural and unattainable.

Now the relation between the boundary classification and the stability concepts introduced in the previous three sections is given. Details are left as exercises. Now besides the assumptions on (4.1) mentioned at the beginning of this section, including (4.2), assume

$$f(0) = g(0) = 0.$$

(4.12)

For any $x > 0$, regularity of the solution

$$X(t) = X(t; 0, x)$$

(4.13)

of (4.1) corresponds to ∞ being unattainable; (4.12) and uniqueness imply that 0 is unattainable, so $X(t)$ takes values in $R_+^o = (0, \infty)$. The trivial solution $X \equiv 0$ is stochastically asymptotically stable (from the right) if and only if 0 is an attracting boundary. If 0 is an attracting boundary and ∞ is a natural boundary, then $X \equiv 0$ is stochastically asymptotically stable in the large (from the right). Noting the first remark in Section 5.3, a Lyapunov function, here, is provided by

$$\bar{\psi}(x) = \int_0^x \exp\left\{-\int_{x_0}^y \frac{2f(z)}{g^2(z)}\, dz\right\} dy.$$

(4.14)

EXAMPLE 5.5   Consider the linear equation

$$dX = aX\, dt + bX\, dW,$$

(4.15)

with a and b positive constants. The function φ in (4.7) has the form here

$$\varphi(x) = \left(\frac{x_0}{x}\right)^{2a/b^2}.$$

So ∞ is a natural boundary if

$$a \le \frac{b^2}{2}$$

(4.16)

since φ is not integrable over intervals of the form (x_0', ∞), $x_0' > 0$, in this case. Further, if

$$a < \frac{b^2}{2}, \tag{4.17}$$

φ is integrable in a neighborhood of 0; thus 0 is attracting.

The function $\bar{\psi}$, given by (4.14), has the form

$$C x^{1-2a/b^2}, \qquad C \text{ a positive constant,}$$

here, and serves as a Lyapunov function verifying stochastic aymptotic stability of the $X \equiv 0$ solution (from the right) when (4.17) holds. Of course, all of these facts are evident from the explicit form of the solution

$$X(t) = X(0) \exp\left(\left(a - \frac{b^2}{2} \right) t + bW(t) \right).$$

Assume now that there is a $K > 0$ such that

$$f(K) = g(K) = 0. \tag{4.18}$$

Under the other assumptions on f and g given above, one can carry out the boundary classification on the intervals $(0, K)$ and (K, ∞). In particular, if 0 and ∞ are natural boundaries, and K is an attracting unattainable boundary (for each interval), then the trivial solution $X \equiv K$ is stochastically asymptotically stable in the large (over $(0, \infty)$). On the other hand, if

$$g(x) \neq 0, \qquad x > 0,$$

(4.1) has no positive constant solutions, and each point in $(0, \infty)$ is attainable from any other. In this case, however, there may exist a stationary distribution. The density p for such a distribution defined by the equation

$$p(y) = \int_0^\infty p(x)p(t, x, y)\, dx,$$

where $p(t, x, y)$ is the transition density associated with the solution of (4.1), must satisfy the time-independent forward equation

$$0 = \frac{d^2}{dx^2}\left(p(x)\tfrac{1}{2}g^2(x) \right) - \frac{d}{dx}(p(x)f(x)) \tag{4.19}$$

and also (to be a density) $p(x) \geq 0$, and

$$\int_0^\infty p(x)\, dx = 1. \tag{4.20}$$

If the function $[g^2(x)\varphi(x)]^{-1}$ is integrable on $(0,\infty)$, it is easy to check that the function

$$p(x) = [Cg^2(x)\varphi(x)]^{-1}, \tag{4.21}$$

where $C = \int_0^\infty [g^2(x)\varphi(x)]^{-1}\, dx$ satisfies (4.19) and (4.20).

EXAMPLE 5.6 [Polansky (1979)] Consider the equation

$$dX = aX(1 - X/K)\, dt + bX\, dW \tag{4.22}$$

on the interval $(0,\infty)$ where a and b are positive constants. The solution of (4.22) can be explicitly obtained (see Example 4.3); although inaccessibility of the boundaries 0 and ∞ is easily observed from the solution form, more detailed qualitative investigation requires a direct approach. It is not difficult to check that ∞ is a natural boundary; and 0 is a natural boundary if $a \geq b^2/2$. The function

$$[g^2(x)\varphi(x)]^{-1} = C_0 x^{2(a/b^2 - 1)} \exp\left(-\frac{2a}{b^2 K}x\right)$$

where C_0 is a positive constant depending on the choice of x_0, is integrable on $(0,\infty)$ provided $a > b^2/2$. So in this case $p(x)$ given by (4.21) provides a stationary density, corresponding to the gamma distribution with parameters $2a/b^2 - 1$ and $2a/(b^2 K)$; the mean is

$$\mu = \frac{2a/b^2 - 1}{2a/(b^2 K)} = K\left(1 - \frac{b^2}{2a}\right)$$

and the variance is

$$\sigma = \frac{2a/b^2 - 1}{[2a/(b^2 K)\,]^2} = K^2\left(1 - \frac{b^2}{2a}\right)\left(\frac{b^2}{2a}\right).$$

When $a > b^2$, $p(x)$ takes its maximum value at μ. The limiting case of (4.22), taking $b \to 0$, corresponds to the deterministic logistic (or Verhulst) model

$$\frac{dx}{dt} = ax\left(1 - \frac{x}{K}\right),$$

which has the property that if $x_0 = x(0) > 0$,

$$x(t; 0, x_0) \longrightarrow K \qquad \text{as } t \longrightarrow \infty.$$

This model, as well as other stochastic analogues of the logistic model, is discussed in more detail in the next chapter. Although verification of the existence of the stationary distribution by constructing the density is preferable, and in fact simpler, when the forward equation can be solved explicitly, note that Theorem 5.9 could also be applied here. Indeed, a

Lyapunov function satisfying the requirements of the theorem can be constructed as follows. Let x_0, x_1, $x_2 \in (0, \infty)$ such that $x_1 < x_0 < x_2$, and let again

$$C = \int_0^\infty [g^2(x)\varphi(x)]^{-1}\, dx.$$

Define the function V by

$$V(x) = \begin{cases} \int_{x_0}^x \varphi(y) \left[C - \int_{x_0}^y [g^2(z)\varphi(z)]^{-1}\, dz \right] dy, & x > x_2 \\[2ex] \nu(x), & x_1 \le x \le x_2 \\[2ex] \int_{x_0}^x \varphi(y) \left[-C - \int_{x_0}^y [g^2(z)\varphi(z)]^{-1}\, dz \right] dy, & x < x_1 \end{cases}$$

$$(4.23)$$

where ν is any nonnegative function chosen so that V is C^2 on $(0, \infty)$. Then V is a nonnegative C^2 function on $(0, \infty)$, and the assumptions

(a) 0 and ∞ are natural boundaries
(b) $g(x) \ne 0$, $x \in (0, \infty)$
(c) $C < \infty$

are employed to establish the other conditions of Theorem 5.9 for this choice of V and $D_{n_0} = (x_1, x_2)$. In particular, the nondegeneracy condition (i) in Theorem 5.9 follows from (b) since (b) and continuity of g imply that

$$g^2(x) \ge M > 0, \qquad x \in [x_1, x_2],$$

for some constant M. Also required is

$$\underset{x \to 0}{\text{limit}}\, V(x) = \underset{x \to \infty}{\text{limit}}\, V(x) = \infty; \qquad (4.24)$$

letting

$$C_1 = \int_0^{x_0} [g^2(z)\varphi(z)]^{-1}\, dz$$

and

$$C_2 = C - C_1 = \int_{x_0}^\infty [g^2(x)\varphi(x)]^{-1}\, dx,$$

from (4.23) it follows that

$$\operatorname*{limit}_{x\to\infty} V(x) \geq C_1 \int_0^{x_0} \varphi(y)\, dy$$

and   (4.25)

$$\operatorname*{limit}_{x\to 0} V(x) \geq C_2 \int_{x_0}^{\infty} \varphi(y)\, dy;$$

assumption (a) implies that the integrals on the right in (4.25) diverge, and so (4.24) is established. Finally, it is an easy calculation to verify that

$$LV(x) = -1, \qquad x \in (0,\infty) \setminus [x_1, x_2],$$

so that (ii) of Theorem 5.9 holds.

5.5 A UNIQUENESS RESULT

Differential inequality techniques apply to problems other than stability. Uniqueness for solutions of initial value problems

$$x'(t) = f(t, x(t))$$
$$x(t_0) = x_0 \tag{5.1}$$

is guaranteed if f satisfies a local Lipschitz condition, but there are several weaker criteria for uniqueness which can be obtained by constructing a Lyapunov-like auxiliary function and establishing a differential inequality. Apparently the first such general theorem is due to Okamura (1942). A number of generalizations have appeared in the literature recently. The main idea of these results is exhibited in the next theorem.

THEOREM 5.10 Let $V(t,x,y)$ be a nonnegative function with continuous first partial derivatives for $t \in [t_0, T]$, $x, y \in R^n$. Assume $V(t,x,y) = 0$ if and only if $x = y$. If

$$\dot{V} = \frac{\partial V}{\partial t} + \frac{\partial V}{\partial x} \cdot f(t,x) + \frac{\partial V}{\partial y} \cdot f(t,y) \leq 0 \tag{5.2}$$

for $t \in (t_0, T)$, $x, y \in R^n$, then there exists at most one solution to the initial value problem (5.1) on the interval $[t_0, T]$.

Note that the condition (5.2) can be written

$$\dot{V} = \frac{\partial V}{\partial t} + \frac{\partial V}{\partial x} \cdot [f(t,x) - f(t,y)] \leq 0, \tag{5.3}$$

when $V(t, x, y) = V(t, x - y)$. The proof of Theorem 5.10 is transparent: $v(t) = V(t, x(t), y(t))$ is a nonnegative nonincreasing function (by (5.2)) which starts out at zero if $x(t)$ and $y(t)$ are any solutions of the initial value problem; thus $v(t) \equiv 0$ on $[t_0, T]$, which implies $x(t) \equiv y(t)$ on $[t_0, T]$. $\square$

EXAMPLE 5.7   Suppose f satisfies the one-sided Lipschitz condition

$$(x - y) \cdot (f(t, x) - f(t, y)) \leq K |x - y|^2 \qquad (5.4)$$

for all $x, y \in R^n$, $t \in (t_0, T)$. Then there exists at most one solution to the initial value problem (5.1). Let $V(t, x, y) = \frac{1}{2} |x - y|^2 e^{-2Kt}$. The left side of (5.3) becomes

$$[-K |x - y|^2 + (x - y) \cdot (f(t, x) - f(t, y))] e^{-2Kt},$$

which is nonpositive by (5.4); so the result follows from the theorem.

Pathwise uniqueness for solutions of stochastic differential equations asserts that the sample paths of solutions of the stochastic initial value problem are identical with probability 1. That is, if $X(t)$ and $Y(t)$ are solutions of the stochastic equation

$$dZ(t) = f(t, Z(t)) \, dt + G(t, Z(t)) \, dW(t) \qquad (5.5)$$

on the interval $[t_0, T]$, and $X(t_0) = Y(t_0)$, then

$$P \left(\sup_{[t_0, T]} |X(t) - Y(t)| = 0 \right) = 1. \qquad (5.6)$$

That pathwise uniqueness is stronger than uniqueness in the law sense (solutions have the same distributions) was proved by Yamada and Watanabe (1971). Pathwise uniqueness is a natural counterpart to uniqueness of solutions of initial value problems involving ordinary differential equations. In Chapter 3 it was proved that Lipschitz conditions on f and G suffice to guarantee pathwise uniqueness. That the Lipschitz condition on f can be weakened to a one-sided condition, analogous to the deterministic example above, follows from the next stochastic Lyapunov-like function theorem.

THEOREM 5.11   Assume

(i)   There is a continuous function $V(t, x, y)$ satisfying

$$a(|x - y|) \leq V(t, x, y) \leq b(|x - y|)$$

$$\text{for } t \in [t_0, T] \text{ and } x, y \in R^n$$

where the functions a and b are continuous and positive definite.

(ii) $LV(t, x, y) \leq 0,$ for $t \in (t_0, T)$ and $x, y, \in R^n$

where L is given by (1.8) with f and G replaced by $f(t, x) - f(t, y)$ and $G(t, x) - G(t, y)$, respectively, that is,

$$LV(t, x, y) = \frac{\partial V}{\partial t} + \sum_{i=1}^{n} [f_i(t, x) - f_i(t, y)] \frac{\partial V}{\partial x_i}$$

$$+ \frac{1}{2} \sum_{i,j=1}^{n} \left(\sum_{k=1}^{m} [g_{ik}(t, x) - g_{ik}(t, y)][g_{jk}(t, x) - g_{jk}(t, y)] \right) \frac{\partial^2 V}{\partial x_i \partial x_j}.$$

Then solutions of (5.5) have the pathwise uniqueness property.

Proof: Making use of the supermartingale inequality (which follows from (ii)), as in the proof of Theorem 5.2, one arrives at

$$P \left(\sup_{[t_0, T]} V(t, X(t) - Y(t)) > \alpha \right) \leq \frac{1}{\alpha} V(t, X(t_0) - Y(t_0)) \qquad (5.7)$$

for any $\alpha > 0$ and solutions $X(t)$ and $Y(t)$ of (5.5). If $X(t_0) = Y(t_0)$, the right side of (5.7) is zero by (i); since α can be taken arbitrarily small, (5.7) leads to

$$P \left(\sup_{[t_0, T]} V(t, X(t) - Y(t)) > 0 \right) = 0. \qquad (5.8)$$

Taking into account (i), (5.8) implies (5.6), which completes the proof.   □

The specific result given in the next example is due to Conway (1971).

EXAMPLE 5.8   Assume that there is a constant K such that

$$(x - y) \cdot (f(t, x) - f(t, y)) \leq K|x - y|^2$$
$$|G(t, x) - G(t, y)|^2 \leq K|x - y|^2$$

$$\text{for all } t \in (t_0, T) \text{ and } x, y \in R^n. \quad (5.9)$$

Then (5.5) has the pathwise unqiueness property.
   Taking

$$V(t, x, y) = \tfrac{1}{2}|x - y|^2 \{\exp -(mn + 2)Kt\},$$

one computes

$$LV(t,x,y) = \left\{ \left(-K - \frac{mnK}{2} \right) |x-y|^2 \right.$$

$$+ \sum_{i=1}^{n} [f_i(t,x) - f_i(t,y)][x_i - y_i] \qquad (5.10)$$

$$\left. + \frac{1}{2} \sum_{i=1}^{n} \sum_{j=1}^{m} [g_{ij}(t,x) - g_{ij}(t,y)]^2 \right\} \exp\{-(mn+2)Kt\}.$$

V clearly satisfies (i) in the theorem, and using (5.9) in (5.10) verifies that
(ii) holds as well. The result follows, then, from the theorem.

EXERCISES

5.1 (a)   Verify that $(0,\infty)$-invariance holds for solutions of the ordinary
           differential equation

$$\frac{dx}{dt} = rx\left(1 - \frac{x}{K}\right)$$

           where r and K are positive constants.

     (b)   Use Theorem 5.3 to show $(0,\infty)$-invariance of solutions of the
           stochastic differential equation

$$dX = rX\left(1 - \frac{X}{K}\right) dt + gX\,dW$$

           where r, K, and g are positive constants, and $r > \frac{1}{2}g^2$. [Hint:
           consider $V(x) = x - K - K\ln(X/K)$.]

5.2 Verify 1.13.

5.3 (a)   Show that solutions of the ordinary differential equation system
           $dx/dt = f(x)$ are bounded if $x \cdot f(x) \le 0$, $|x| \ge K$, some constant
           $K > 0$.

     (b)   Show that the condition given in part a) also yields stochastic
           sample path boundedness up to first exit from $\{|x| > K\}$ for the
           system

$$dX = f(X)\,dt + G(X)\,dW$$

           where the matrix function $G(x)$ has the form

$$G(x) = g(x)I,$$

$g(x)$ a scalar function and I the identity matrix.

5.4 (a) Prove that stochastic sample path boundedness for the scalar equation

$$dX = f(X)\,dt + g(X)\,dW$$

with $g(x) \neq 0$, $x \in R$, implies $\pm\infty$ are natural boundaries.

(b) Assume $f(0) = g(0) = 0$. Prove that the trivial solution $X \equiv 0$ of the stochastic equation in part (a) is stochastically asymptotically stable if and only if 0 is an unattainable attracting boundary for the solution process on $(-\infty, 0)$ and $(0, \infty)$.

(c) Verify that the asymptotic stability obtained in part b occurs in the large if, in addition, $\pm\infty$ are natural boundaries.

5.5 Prove the second part of Theorem 5.7. [Hint: Observe that the proof given in the text now holds for arbitrary $r > 0$. Apply Theorem 5.2 to obtain stochastic sample path boundedness; what can be said about the number of paths that exit every B_r in finite time?]

5.6 Assume (3.4) holds, where A is a symmetric negative definite matrix. Recall, from Example 5.2 in Section 5.2, that the trivial solution $x \equiv 0$ of the O.D.E. $dx/dt = f(t,x)$ is asymptotically stable in this case. Show that the trivial solution $X \equiv 0$ of (3.3) is stochastically asymptotically stable also in this case, if, for each k, B_k is the diagonal matrix with lone nonzero entry b in the kth location [Hint: Note that in (3.6), in this case, $B(y) = bIy$, where I is the identity.]

5.7 Prove Theorem 5.8 [Hint: For $\delta < |x| < r$, let $\tau_{\delta,r}(t) = \min\{t$, first hitting time of the set $\{|x| = \delta\} \cup \{|x| = r\}\}$ and $\tau_\delta =$ first hitting time of $\{|x| = \delta\}$; show that $\tau_\delta \to \infty$ as $\delta \to 0$ w.p. 1, and $X(t) = X(t; s, x)$ implies

$$\inf_{|x| < \varepsilon, t > 0} V(t, x) P\left(\sup_{[s, \tau_\delta]} |X(t)| < r\right) < EV\left(\tau_{\delta,r}(t), X(\tau_{\delta,r}(t))\right).$$

Apply the supermartingale inequality.]

5.8 Consider the system

$$dX = [Y - Xf(X,Y)]\,dt + g_1 X\,dW_1$$

$$dY = [-X + Yf(X,Y)]\,dt + g_2 Y\,dW_2,$$

where W_1 and W_2 are independent scalar Wiener processes, f is a continuous function, and g_1 and g_2 are constants. Show that the trivial solution is stochastically asymptotically stable provided $f(x,y) \geq (1/2)\max\{g_1^2, g_2^2\}$. [Hint: Use the Lyapunov function $V = (1/2)(x^2 + y^2)$. Note that $Q(x)$ is indefinite, in this case, so that the linearization result does not apply.]

5.9   (Has'minskii)   The trivial solution $X(t) \equiv 0$ of the system (1.6) is *exponentially p-stable* [exponentially stable in the mean (mean square) for $p = 1(2)$] if for some positive constants K and k

$$E|X(t; s, x)|^P \leq K|x|^P \exp\{-k(t-s)\}.$$

Suppose there is a twice differentiable function $V(t, x)$ satisfying

(i) $a|x|^P \leq V(t, x) \leq b|x|^P.$
(ii) $LV(t, x) \leq -c|x|^P.$

for all $t \geq 0$ and $x \in R^n$, where a, b, and c are positive constants. Prove that the trivial solution of (1.6) is exponentially p-stable.

5.10 Determine the boundary classification for the diffusion process defined by the Ito equation

$$dX = X^\alpha \, dt + X^\beta \, dW$$

on the interval $(0, \infty)$. Does an invariant distribution exist?

6

Applications in Biology:
Population Dynamics

6.0   INTRODUCTION

Recently, significant applications of mathematics have been extended to problems in the social and life sciences. The need and desire to solve difficult basic problems, such as disease control, which affect the quality of life, have spurred progress in all the sciences, including mathematics. For many of these problems the usual experimental methods fundamental to scientific research are expensive, if not impossible, to implement. Mathematical modeling and subsequent analysis have proved productive in a number of these cases, either replacing or augmenting empirical techniques. Applications of mathematics relating to the life sciences extend from the microscopic scale as in genetics or neurophysiological problems to the macroscopic level as in the environmental questions of pollution, renewable resource management, and population dynamics; mixed scale problems occur in population genetics and epidemiology, for example.   The need to take into account random phenomena explicitly to achieve desired accuracy in a number of these modeling situations has motivated the use of stochastic process models in particular.   The random aspects considered may be intrinsic, for example, the firing rate of a neuron, or external, such as random environmental characteristics affecting a population community, or both, as in ecological population genetics problems. Markov diffusion processes, which can be represented as solutions of stochastic differential equations, arise as tractable approximations to key stochastic model quantities, such as gene frequency, neural membrane electrical potential, and population density. In this chapter attention is focused on some aspects of recent applications of stochastic differential equations to population dynamics. In Section 2 some simple single population models are discussed, followed by extensions to the community level in Section 3. To begin, in the next section, some well-known basic limit theorems are given to illustrate, in a precise way, how stochastic differential equation models may arise.

6.1   MODELING CONSIDERATIONS AND APPROXIMATIONS

In many cases a model of a physical system is specified by a deterministic function, for example, the position of a moving particle as a function of time. From a description (such as the equation of motion of a particle) of the model, the problem is to obtain qualitative information about the function, if not to exhibit it in some explicit closed form (for example, a Fourier series). If random effects are to be considered, the appropriate model is a stochastic process, which can be thought of as a family of functions indexed by a probability space parameter. The probability of a given event in this parameter space measures how likely functions in that event represent the system. The functions, or sample paths of the process, may be described themselves by a stochastic equation, or more generally the probability law of the process may be given by the solution of some master equation, such as the Chapman-Kolmogorov equation. As is sometimes the case even for deterministic models, the describing equations for the stochastic model may be mathematically intractable. Hence further simplification is necessary before a successful analysis can be executed. Available data together with some understanding of the mechanism being modeled may permit one to conclude that the model should belong to some special class of processes, or at least be approximated by processes in some such class. For systems in which differential and difference equations have served well in the deterministic situation, the class of Markov diffusion processes is a natural choice in the random case. Tractability of such processes characterized as solutions of stochastic differential equations has been illustrated in this book to some extent. The limit theorems given in this section illustrate that stochastic differential equations may arise as approximations to possibly more physically intuitive but less tractable mathematical models; as such, the results mentioned here serve two purposes: (1) to give an indication why diffusion processses are a natural choice, and (2) to present some basis for deciding whether the Ito or the Stratonovich (or some other) approach provides the appropriate diffusion model or approximation. It is emphasized that the Ito and Stratonovich interpretations of stochastic differential equations provide only two of a number of methods for implementing diffusion approximations.

The applications to be considered are from biology, population dynamics in particular. Thus state variables $X(t)$ represent biological components, for example, population densities, at time t, and the equations for X model how these components evolve in time. The limit theorems discussed in this section are interpreted within this framework; the proofs, which involve a number of technical details, are omitted to maintain continuity of the presentation and focus on the modeling problem. Also, for simplicity, in this section attention is restricted to autonomous one-dimensional (single-component) models. The discussion essentially follows Turelli (1977).

Suppose, first of all, that an ordinary differential equation perturbed by a correlated noise process provides an accurate model of a particular physical system operating in a random environment. That is, the basic model quantity $X(t)$ is assumed to satisfy an equation of the form

$$\frac{dX}{dt} = f(X) + g(X)\xi(t) \tag{1.1}$$

where $\xi(t)$ is some normalized autocorrelated noise process. As mentioned previously, for similar equations, the function $f(x)$ represents the dynamics of the system in the absence of random effects, while $g(x)$ indicates the intensity of the noise. Also, it has been noted that the nature of the noise process $\xi(t)$ often makes rigorous mathematical interpretation, let alone analysis, of (1.1) difficult. Toward obtaining a more mathematically tractable model and clarifying the underlying model assumptions, (1.1) may be replaced by the sequence of equations

$$\frac{dX_n}{dt} = f(X_n) + g(X_n)\xi_n(t) \tag{1.2}$$

where the ξ_n are processes similar to ξ which also exhibit some specific limiting behavior as n becomes large. The idea is to establish for (1.2) that, in a sense that is consistent with the asymptotic behavior of ξ_n,

$$X_n \longrightarrow X_L \qquad \text{as } n \longrightarrow \infty \tag{1.3}$$

where X_L is a process whose mathematical properties can be more readily determined than those of X in (1.1). Thus, for large n, X_L approximates X_n—its properties are somehow similar to those of the more immediate but less tractable processes X_n. This situation is made precise when the sense in which (1.3) holds is specified, and the theorems stated in this section conclude exactly that. [The third theorem mentioned, however, involves difference rather than differential equations corresponding to (1.1) and (1.2).]

Generally, the processes ξ_n in (1.2) are assumed to be strictly stationary with mean zero and covariance function

$$K_n(s) = E(\xi_n(t+s)\xi_n(t)), \tag{1.4}$$

and approach Gaussian white noise in some sense, as n becomes large. Precisely how ξ_n tends to white noise determines two types of theorems which obtain X_L in (1.3) as a Markov diffusion process.

In one type (Stratonovich, 1966; Wong and Zakai, 1965b), the ξ_n are taken as smooth sample path approximations to Gaussian white noise as the "derivative" of the Wiener process W. The theorem of Wong and Zakai (1965b), given below, establishes X_L as the Stratonovich solution of the stochastic differential equation

$$dX = f(X)\,dt + g(X)\,dW \tag{1.5}$$

It requires the following additional conditions on ξ_n:

$$\xi_n = \frac{dY_n}{dt} \tag{1.6}$$

where, with probability 1, the sample paths of Y_n are continuous and of bounded variation with piecewise continuous derivatives and converge to the paths of W as $n \to \infty$; also there exist a random integer n_0 and a random variable k, both with finite mean, such that $Y_n(t) \leq k$, all $n \geq 0$, $t \in [a,b]$. (1.7)

THEOREM 6.1   (Wong and Zakai, 1965b)   Assume that, for all $x, y \in R$,

(i)   The functions f, g, and g' satisfy the Lipschitz condition for some constant k_1,

$$|f(x) - f(y)| + |g(x) - g(y)| + |[gg'](x) - [gg'](y)| \leq k_1|x - y|.$$

(ii)   For some constant k_2, $g^{-2}(X) \leq k_2$.

Suppose further that $X_n(t)$ and $X(t)$ are solutions of (1.2) and (1.5) (Stratonovich sense), respectively, on a interval $[a,b]$ and

$$X_n(a) = X_a = X(a)$$

where X_a is a random variable independent of the differences $W(t) - W(a)$, all $t \in [a,b]$. If the ξ_n satisfy (1.6), (1.7), then

$$X_n(t) \longrightarrow X(t) \qquad \text{as } n \longrightarrow \infty, \qquad \text{w.p. 1,} \quad t \in [a,b].$$

A second type of theorem (Has'minskii, 1980; Papanicolaou and Kohler, 1974), which can be considered as a generalization of Theorem 6.1, results in a one-parameter family X_γ of Markov diffusion processes in the role of X_L. Convergence of X_n to X_γ occurs in distribution (or the weak sense). In this case the ξ_n provide probability law approximations to white noise; they are determined by scaling some smoothly autocorrelated process η with the parameter γ indicating the strength of the correlation. The comparable case of Papanicolaou and Kohler's result follows the precise conditions required on ξ_n:

$$\xi_n(t) = \sqrt{n}\,\eta(nt) \tag{1.8}$$

where η is a continuous, strictly stationary, mean zero process with covariance function K and satisfying the mixing condition: there is a function φ such that for any events A and B, any t, and $s > 0$

$$|P(\eta(t+s) \in A | \eta(t) \in B) - P(\eta(t+s) \in A)| \leq \varphi(s)$$

$$\text{and } \varphi(s) \longrightarrow 0 \qquad \text{as } s \longrightarrow \infty. \tag{1.9}$$

Finally, set

$$\gamma = \int_{-\infty}^{\infty} K(s)\,ds. \tag{1.10}$$

THEOREM 6.2   (Papanicolaou and Kohler, 1974) Assume that, for all $x \in R$, the function f satisfies, for some constant C and integer $q \geq 0$,

(i) $|f(x)| \leq C(1 + |x|)$,
(ii) $|f'(x)| \leq C$,
(iii) $|f''(x)| + |f'''(x)| + |f^{(4)}(x)| \leq C(1 + |x|^q)$,

and also that the function g satisfies (i) to (iii). Suppose further that the ξ_n processes in (1.2) satisfy (1.8) to (1.10), with

$$\int_0^{\infty} \sqrt{\varphi(s)}\,ds < \infty. \tag{1.11}$$

Let $X_n(t)$ be a solution of (1.2) on some interval $[a, b]$ satisfying $X_n(a) = X_a$, a random variable. Then, for $t \in [a, b]$,

$$X_n(t) \longrightarrow X_\gamma(t) \qquad \text{as } n \longrightarrow \infty, \text{ in the distribution sense}$$

where X_γ is the diffusion process on $[a, b]$ satisfying $X_\gamma(a) = X_a$ and having drift coefficient $f(x) + \gamma g(x)g'(x)$ and diffusion coefficient $\gamma g^2(x)$.

The scaling (1.8) of η to obtain ξ_n makes the passage of a small amount of time s for ξ_n correspond to a large time ns for η and increases the variance by a factor of n at each t. Condition (1.11) implies that

$$\gamma = \int_{-\infty}^{\infty} K(s)\, ds = \int_{-\infty}^{\infty} K_n(s)\, ds \tag{1.12}$$

is finite. By invoking the central limit theorem, it can be seen that $\xi_n/\sqrt{\gamma}$ tends to standard Gaussian white noise in the distribution sense. Only when $\gamma = 1$ holds does the limiting process X in Theorem 6.2 coincide with the Stratonovich solution of (1.5); the Ito solution of (1.5) never arises as a special case of this result. In general, then, Theorem 6.2 demonstrates that although (1.2) may be very close to a white noise forced differential equation, the appropriate diffusion approximation may be supplied by neither the Ito nor the Stratonovich-associated stochastic differential equation.

In some sense, autocorrelation is more essentially involved in the construction of the noise via (1.8) to (1.10) rather than (1.6), (1.7). Indeed, in the former case it is easy to see that

$$\int_0^t \xi_n(s)\, ds = \sqrt{n} \int_0^t \eta(ns)\, ds = \frac{1}{\sqrt{n}} \int_0^{nt} \eta(u)\, du$$

$$\longrightarrow \sqrt{\gamma} W(t), \qquad \text{in the weak sense} \tag{1.13}$$

while, for the latter,

$$\int_0^t \xi_n(s)\, ds = \int_0^t \frac{dY_n(s)}{ds}\, ds = Y_n(t) \longrightarrow W(t), \qquad \text{w.p.1,} \tag{1.14}$$

as $n \to \infty$. This situation is evidenced in the preservation of some aspect of correlation in the parameter γ of the limiting proccess in Theorem 6.2.

The significance of Theorem 6.1 or 6.2 for applications, of course, depends on whether for large n (1.2) with either (1.6), (1.7) or (1.8) to (1.10) constitutes an acceptable model for the particular physical system of interest. In corresponding mechanics or engineering problems, for example, the variable X usually represents a generalized position coordinate of an object moving in an environment subject to random disturbances. In the absence of random effects, the dynamics of X are determined by established equations of motion which take the form of differential equations. In this case a process $X(t)$ with smooth (at least differentiable) sample paths whose rate of change is subject to correlated noise perturbations is often an intuitive and justifiable model type—hence, the applicability of Theorems 6.1 and 6.2, as argued in much of the Ito versus Stratonovich controversy literature. However, in population dynamics, for example, differential equation models arise mainly as approximations themselves to more relevant discrete models. (Population densities neither are measured continuously nor vary continuously.) Factors in natural population dynamics such as nonoverlapping generations, discrete breeding seasons, and environmental parameters exhibiting significant change only at discrete times argue for discrete-type models such as difference equations. Since difference equations, however, are difficult to analyze generally, these models are replaced by more tractable ones having similar qualitative features. For population dynamics in random environments, then, seeking diffusion approximations for appropriate stochastic difference equation models seems to constitute a logical approach to this problem.

It has already been observed in this book that diffusion processes arise as limits of the solutions of difference equations of the form

$$X_n\left(t + \frac{1}{n}\right) - X_n(t) = f(X_n(t))\frac{1}{n} + g\left(\lambda X_n\left(t + \frac{1}{n}\right)\right.$$

$$\left. + (1 - \lambda)X_n(t)\right)\left(W\left(t + \frac{1}{n}\right) - W(t)\right)$$

$$\text{for } t = 0, \frac{1}{n}, \frac{2}{n}, \dots \text{ and } \lambda \in [0,1]. \quad (1.15)$$

Indeed, Section 3.4 indicates that, under suitable conditions on f and g, $X_n(t) \to X_\lambda(t)$, where $X_\lambda(t)$ is a Markov diffusion process with drift coefficient $f(x) + \lambda g(x)g'(x)$ and diffusion coefficient $g^2(x)$. In particular, the Stratonovich solution of (1.5) is not excluded as a possible diffusion approximation for corresponding sequences of difference equations (the $\lambda = 1$ case). On the other hand, an aspect of (1.15) that makes this model questionable in the population dynamics context is the dependence of the noise intensity term

$$g\left(\lambda X_n\left(t + \frac{1}{n}\right) + (1 - \lambda)X_n(t)\right)$$

on the future variable $X_n(t + 1/n)$ if $\lambda > 0$. Eliminating this type of dependence in (1.15) excludes the possibility of Stratonovich's interpretation of (1.5) arising. Also, aspects contributing to randomness in the environment such as food supply and weather variability indicate that accurate models ought to include autocorrelated noise. Diffusion approximations for such stochastic difference equations involving autocorrelated noise with nonanticipating intensity have been obtained (Skorohod, 1958; Gihman and Skorohod, 1969; Kushner, 1974) as the Ito solution to (1.5).

Whether starting from difference or differential equations, any of these methods for obtaining a diffusion processs model requires selecting functional forms for the average unit or rate of change $f(x)$ and the noise intensity $g(x)$ of the random environment. Experimental data give insight into the choice, but usually some a priori heuristic guess at a general form for f and g must be made, and then validated by fitting the data. A typical procedure, discussed briefly earlier in this book, begins with a deterministic model, continuous

$$\frac{dx}{dt} = f_0(x) \tag{1.16}$$

or discrete

$$x_{n+1} - x_n = f_0(x_n), \tag{1.17}$$

in which one has some confidence. A parameter λ involved in the functional form of f_0 which is likely to be affected by random environmental variations is written as the sum

$$\lambda = \alpha + \sigma \tag{1.18}$$

of an average α and a fluctuation term σ. If the parameter appears linearly in f,

$$f_0(x) = f_1(x) + \lambda f_2(x), \tag{1.19}$$

a natural choice for f and g in a stochastic counterpart of either (1.16) or (1.17) would be

$$f(x) = f_1(x) + \alpha f_2(x)$$
$$g(x) = \sigma f_2(x). \tag{1.20}$$

A danger of this heuristic procedure is that naturally restricted deterministic parameters may be replaced by analogous unrestricted behavior in the stochastic model. For example, stochastic difference equations

$$x_{n+1} - x_n = f(x_n) + g(x_n)\xi_n$$

derived from (1.17) via (1.18) to (1.20), where ξ_n is some normalized approximate noise process, may predict negative or arbitrarily large values for x_n—a very inappropriate situation for population density models.

Turelli's extension (1977) of Kushner's result, given below, deals with this situation by replacing f and g in (1.21) by smooth bounded truncations f_n and g_n; f_n and g_n agree with f and g for $x \leq C_n$, $C_n \to \infty$, as $n \to \infty$ and for $x > C_n$, f_n and g_n may be defined arbitrarily so that unrealistic values for x_n are eliminated. Successful execution of the program outlined (i.e., validation of the difference equation sequence) by Turelli's theorem argues for the applicability, in this situation, of diffusion approximations determined by the functions f and g, that is, solutions of the Ito-interpreted stochastic differential equation (1.5) where

> the functions f and g are continuously differentiable on $[0, \infty)$
> with $f(0) = g(0) = 0$ (1.21)

and

> for any solution of (1.5) on $[0, \infty)$, the boundary at ∞ is
> unattainable.   (1.22)

Conditions (1.21) and (1.22) guarantee that if X_a is a random variable
independent of the Wiener differences $W(t) - W(a)$, $t \geq a$, the stochastic
initial value problem consisting of the Ito-interpreted equation (1.5) and
initial value

$$X(a) = X_a \tag{1.23}$$

has a unique solution $X(t)$ taking values in $[0, \infty)$ and defined for t in the
interval $[a, \tau)$ where

$$\tau = \infty \qquad \text{or} \qquad X(\tau) = 0;$$

furthermore, by uniqueness, if $X_a = 0$ w.p. 1,

$$X(t) \equiv 0 \qquad t \in [a, \infty), \text{ w.p. } 1.$$

The theorem stated below relates $X(t)$ to the solutions $X_N(t)$ of initial
value problems involving stochastic difference equations of the form

$$X_N(a + n + 1) - X_N(a + n)$$

$$= f_N(X_N(a + n)) \frac{1}{N} + g_N(X_N(a + n)) \frac{\eta_N(a + n + 1)}{N^{1/2}} \tag{1.24}$$

and initial value

$$X_N(a) = X_a. \tag{1.25}$$

In (1.24), it is asssumed that

> for each N, the functions f_N and g_N are bounded con-
> tinuously differentiable functions on $[0, \infty)$, and there are
> numbers $C_N, C_N \to \infty$ as $N \to \infty$, such that
>
> $$f_N(x) = f(x), g_N(x) = g(x), x \leq C_N, \tag{1.26}$$
>
> and that each η_N is sequence of i.i.d. mean zero, variance
> one random variables with finite third moments.   (1.27)

A solution process X_N of (1.24), (1.25), whose existence is guaranteed
by basic difference equation theory, is extended to a continuous time process

(with right continuous sample paths) by defining

$$X_N(t) = X_N(a+n), \qquad a+n \le t < a+n+1. \tag{1.28}$$

THEOREM 6.3   (Turelli, 1977)   Assume f and g are functions satisfying (1.21), and suppose (1.22) holds. Let $X_N(t)$ be constructed as in (1.24) to (1.28). If $X(t)$ is the Ito solution of the stochastic initial value problem (1.5), (1.23), then, in the weak sense,

$$X_N(t) \longrightarrow X(t) \qquad \text{as } N \longrightarrow \infty.$$

For example, the stochastic analogues

$$dX = aX\,dt + bX\,dW \tag{1.29}$$

and

$$dX = a(1 - X/K)X\,dt + g(X)\,dW \tag{1.30}$$

of the basic deterministic linear and logistic models respectively can be constructed as population dynamics models via Theorem 6.3. Solutions of initial value problems involving (1.29) and (1.30) with $g(x) = bx$ respectively can be obtained explicitly in terms of W:

$$X(t) = X(0) \exp\left\{ (a - b^2/2)t + bW(t) \right\}$$

and

$$X(t) = \frac{X(0) \exp\left\{ (a - b^2/2)t + bW(t) \right\}}{1 + \dfrac{X(0)}{K} a \displaystyle\int_0^t \exp\left\{ (a - b^2/2)s + bW(s) \right\}\,ds}$$

for all $t \ge 0$. The boundary condition (1.22) is obvious from these forms and the sample path properties of $W(t)$. (These and other specific examples are treated in more detail in the next section.)

Suppose, more generally, that g is a continuous function on $[0, \infty)$, continuously differentiable on $(0, \infty)$, and

$$g(0) = 0, \quad g(x) \ne 0, \quad x \ne 0. \tag{1.31}$$

Then it follows that (1.22) holds for (1.30). To see this note first that

$$ax(1 - x/K) \le 0 \qquad \text{for } |x| \ge K$$

implies that

$$\varphi(x) = \exp\left\{ -\int_K^x \frac{2f(y)}{g^2(y)}\,dy \right\} \ge 1, \qquad \text{for } x \ge K.$$

Hence φ is not integrable on the interval (K, ∞), and so by the boundary classification scheme of Feller (Section 5.4) ∞ is a natural boundary, hence unattainable. This remark indicates the general applicability of Theorem 6.3 to density dependent population models as summarized in the following corollary.

COROLLARY 6.4 Assume f and g are functions satisfying (1.21) and (1.31), and suppose, for some $K > 0$,

$$f(x) \leq 0 \qquad \text{for } x \geq K.$$

Then the conclusion of Theorem 6.3 holds.

The limit theorems of this section indicate that similar deterministic equations, viewed as approximations of stochastic models, can lead to different limiting diffusion processes. The controversy [see the articles by Gray and Caughy (1965), Mortensen (1969), Turelli (1977, 1978), and van Kampen (1981), and the books by Saaty (1967), Wong (1971), and McShane (1974)] over the interpretation of (1.5) as an Ito equation or Stratonovich equation often resolves itself under careful derivation of (1.5) from the proper deterministic counterpart. If the state of a physical system can be accurately represented by the solution of a differential equation in the absence of a noise input, such as the charge on a capacitor in an electrical circuit, then the corresponding random state arising from external noise forcing may be represented by the Stratonovich solution of (1.5), arguing along the lines of Theorem 6.1. On the other hand, in some situations, for example, biological population growth in constant environments, difference equations provide more precise deterministic models, although their relative mathematical intractability prompts their replacement with approximate differential equations. To take into account stochastic effects, such as random environments, for such models it is therefore appropriate to start with the difference equation rather than the more remote differential equation. Then results such as Theorem 6.3 indicate that the Ito interpretation of (1.5) may be the appropriate stochastic model. However, whether the "correct" deterministic model is a differential equation or a difference equation is only one of number of considerations in deriving a useful stochastic model. For example, if a parameter, such as the carrying capacity in a biological growth model, which has an intrinsically restricted range is replaced by its mean value plus a noise term, the interpretation of the resulting stochastic equation is unclear (Turelli, 1977). Furthermore, if the source of randomness is not exclusively external to some deterministic system, starting with a deterministic model of any type to obtain the desired stochastic model may be unsound (van Kampen, 1981).

6.2   SOME POPULATION DYNAMICS MODELS

The simplest population dynamics models take the form of the differential equation (1.16) or the difference equation (1.17) with

$$f_0(x) = ax,$$

where a is a (positive) constant, representing the growth rate of the species modeled. The major drawback of such models, of course, is their prediction of unbounded growth. Corresponding stochastic equations which attempt to model random environmental effects on the population exhibit similar behavior. In particular, this situation does not change if, following the heuristics in this last section and earlier in this book, one replaces the growth rate a by the sum

$$\alpha + \beta \mathcal{N}(t),$$

where α and β are constants, and $\mathcal{N}(t)$ is white noise, and proceeds, possibly by invoking a theorem of the last section, to the stochastic differential equation

$$dX = \alpha X \, dt + \beta X \, dW. \tag{2.1}$$

Whether interpreted as an Ito equation, in which case the solution is

$$X_I(t) = X(0) \exp \left\{ (\alpha - \tfrac{1}{2}\beta^2)t + \beta W(t) \right\} \tag{2.2}$$

or a Stratonovich equation, whose solution is

$$X_S(t) = X(0) \exp \left\{ \alpha t + \beta W(t) \right\}, \tag{2.3}$$

the specter of arbitrarily large population densities remains, although at least in the former case the noise intensity restricts this behavior. In fact, since

$$\frac{W(t)}{t} \longrightarrow 0, \qquad \text{w.p. 1 (see Chapter 1)},$$

with probability 1,

$$X_I(t) \longrightarrow 0 \text{ as } t \longrightarrow \infty \text{ if } \mu < \tfrac{1}{2}\sigma^2$$

and $\tag{2.4}$

$$X_I(t) \longrightarrow \infty \text{ as } t \longrightarrow \infty \text{ if } \mu > \tfrac{1}{2}\sigma^2$$

while

$$X_S(t) \longrightarrow 0 \text{ as } t \longrightarrow \infty \text{ if } \mu < 0$$

and   (2.5)

$$X_S(t) \longrightarrow \infty \text{ as } t \longrightarrow \infty \text{ if } \mu > 0.$$

The solutions X_I and X_S constitute diffusion processes on $(0, \infty)$ with common diffusion coefficient

$$g^2(x) = \sigma^2 x^2$$

and drift coefficients μx and $(\mu + \frac{1}{2}\sigma^2)x$ respectively; the boundaries 0 and ∞ are unattainable in both cases, and conditions (2.4) and (2.5) specify which boundary point is attracting in terms of the average growth rate μ (Exercise 6.2).

Environmental limitations as well as self-regulation prevent arbitrarily large densities for natural populations. Deterministic models of single-species population growth which take into account self-regulatory mechanisms have the form (1.16) or (1.17) with

$$f_0(x) = rx(1 - \varphi(x))$$

or, in the nonautonomous case,

$$f_0(t, x) = r(t)x(1 - \varphi(t, x))$$

where φ is an increasing function in x. The simplest such model is the well-known logistic (or Verhulst) model

$$f_0(x) = rx(1 - x/K);$$ (2.6)

here the parameters r and K are positive constants referred to as the population's intrinsic growth rate and carrying capacity, respectively. The differential equation (1.16) with f_0 given by (2.6) exhibits the solution behavior

$$x(t) \longrightarrow K \text{ as } t \longrightarrow \infty \text{ if } x(0) = x_0 > 0$$

monotonically, that is if $x_0 < K$, $x(t) \uparrow K$, all t; for the corresponding difference equation (1.17), a solution for which $x_n > \tilde{K} = K(1 + 1/r)$, some nonnegative integer n, must also satisfy $x_m < 0$, all $m > n$, that is, the model is meaningful only for $x_n \leq \tilde{K}$, all n.

In the remainder of this section, several stochastic analogues of the deterministic logistic model are briefly discussed using the theory developed in Section 5.4. Verification of details are left as exercises. One might begin by replacing r by $\rho + \sigma N(t)$ or $1/K$ by $1/K + \sigma N(t)$, arriving at either of the stochastic equations

$$dX = \rho X(1 - (X/K))\, dt + \sigma X(1 - (X/K))\, dW \tag{2.7}$$

or

$$dX = rX(1 - (X/K))\, dt - \sigma r X^2\, dW. \tag{2.8}$$

Regardless of the choice of Stratonovich or Ito interpretation, these equations represent two extremes of relative preservation of the self-regulatory behavior of the deterministic logistic model. In (2.7), K maintains its carrying capacity role as an unattainable attracting boundary for either Ito or Stratonovich diffusion process solution on the interval $(0, K)$. On the other hand, (2.8) permits solutions to attain arbitrarily large values with positive probabilities. The carrying capacity as an upper bound for population density is either preserved [in (2.7)] or completely destroyed [in (2.8)]. The latter situation also occurs for the model

$$dX = aX(1 - X/K)\, dt + bX\, dW \tag{2.9}$$

discussed in Section 5.4. One may consider (2.9) as arising from the logistic parameterization

$$f_0(x) = x(r - sx) \tag{2.10}$$

by replacing the growth rate r by the usual

$$a + b\mathcal{N}(t);$$

note, however, that this forces the role of the carrying capacity r/s in the deterministic model also to be replaced by an unrestricted stochastic parameter

$$\frac{1}{s}(a + b\mathcal{N}(t))$$

just as in (2.8). Turelli (1977) suggests that a more reasonable approach toward obtaining a stochastic analogue of the logistic model may be to start with Schoener's parametrization

$$f_0(x) = Rx[E(T - \Lambda x) - C - \Gamma x]; \tag{2.11}$$

in (2.11) R is a conversion factor of energy into offspring, T is the proportion of time spent in feeding and intraspecific interaction, E is the net energy obtainable per unit foraging time, and Λ, C, and Γ measure costs of metabolic maintenance and intraspecific interaction. The intrinsic growth rate and carrying capacity for (2.11) are:

$$r = R(ET - C) \quad \text{and} \quad K = (ET - C)/(\Gamma + E\Lambda). \tag{2.12}$$

EXAMPLE 6.1   To model fluctuation resource abundance, which affects both r and K, one might replace E by

$$\varepsilon + \sigma \mathcal{N}(t),$$

which, in contrast to (2.7), (2.8), and (2.9), leads to the stochastic equation

$$dX = r_0 X(1 - (X/K_0))\, dt + \sigma R X(T - \Lambda X)\, dW \qquad (2.13)$$

where r_0 and K_0 are given by (2.12) with E replaced by ε. Note that, in particular,

$$K_0 = \frac{\varepsilon T - c}{\Gamma + \varepsilon \Lambda} < \frac{T}{\Lambda} = K_1. \qquad (2.14)$$

The Ito solution $X(t)$, $t \geq 0$, with $X(0) = x$, $x \in (0, K_1)$, constitutes a diffusion process on $(0, K_1)$ and the boundary behavior can be determined. The facts that

$$g(K_1) = \sigma R K_1 (T - \Lambda K_1) = 0,$$

and, following from (2.14), that

$$f(x) = r_0 x (1 - x/K_0) < 0, \qquad x \text{ near } K_1$$

suggest that K_1 is a repelling, hence unattainable, boundary; this is confirmed by applying the boundary analysis for scalar diffusion processes given in Section 5.4:

1. First, for x near K_1, the function

$$
\begin{aligned}
\varphi(x) &= \exp\left\{ -\int_{x_0}^{x} \frac{2f(y)}{g^2(y)}\, dy \right\} \\[2mm]
&= C x^{-C_0} \left(1 - \frac{x}{K_1}\right)^{C_0} \exp\left\{ C_0 \left(\frac{K_1}{K_0} - 1\right)\left(1 - \frac{x}{K_1}\right)^{-1} \right\}
\end{aligned}
\qquad (2.15)
$$

where C is an arbitrary constant dependent on x_0 and $C_0 = 2r_0/\sigma^2 R^2 T^2$, behaves like $u^p e^{1/u}$, near $u = 0$. The function φ, therefore, is not integrable in a neighborhood of K_1, and so by the boundary classification scheme K_1 is repelling (or natural), in constrast to K for (2.7) to (2.9).

2. From (2.15) it also follows that

$$|\varphi(x)| \approx C_1 x^{-C_0}$$

for x near 0, where C_1 is a constant; so 0 is repelling (and hence unattainable) if $C_0 \geq 1$, and attracting if $C_0 < 1$. In either case the boundary 0 is unattainable also, since

$$\xi(x) = \varphi(x) \int_{x_0}^{x} [g^2(y)\varphi(y)]^{-1}\, dy$$

satisfies, for x near 0, for some constant C_2,

$$|\xi(x)| \approx C_2 \begin{cases} -x^{-1}\ln x, & \text{if } C_0 = 1, \\ x^{-1}, & \text{if } C_0 \neq 1. \end{cases}$$

and so ξ is not integrable in a neighborhood of 0.

3. Furthermore, since

$$[g^2(x)\varphi(x)]^{-1} = C_3 x^{C_0 - 2} \left(1 - \frac{x}{K_1}\right)^{-C_0 - 2}$$

$$\times \exp\left\{-\frac{C_0((K_1/K_0) - 1)}{1 - x/K_1}\right\},$$

for C_3 a positive constant, this function is integrable on $(0, K_1)$ if $C_0 > 1$: it is continuous on $(0, K_1)$, near 0 it behaves like

$$x^{C_0 - 2},$$

while near K_1 it behaves like $u^p e^{-u}$, p a positive number, u near ∞. Thus

$$p(x) = C_4[g^2(x)\varphi(x)]^{-1},$$

$$\text{for } C_4 = \left[\int_0^{K_1} [g^2(x)\varphi(x)]^{-1} \, dx\right]^{-1},$$

is a stationary distribution density in this case. When $C_0 > 2$, the density has a unique maximum; if, in addition $K_1 = 2K_0$, the maximum occurs at K_0.

To summarize, for the stochastic logistic analogue (2.13) obtained from the Schoener parametrization (2.11), the carrying capacity K_0 in the deterministic model is stretched out to the repelling boundary K_1 for the stochastic model and the size of the parameter C_0 determines the relative stability of the stochastic model: if $C_0 < 1$, 0 is an attracting boundary, if $C_0 \geq 1$, 0 is a repelling boundary, and further if $C_0 > 1$, (2.13) has a stable invariant distribution with nowhere zero density in $(0, K_1)$. These facts seem to indicate that (2.13) is a more palatable analogue of the logistic than either (2.7), (2.8), or (2.9).

6.3 MULTISPECIES POPULATION DYNAMICS

6.3.1 Deterministic Models

The Lotka-Volterra system of ordinary differential equations

$$\frac{dx_i}{dt} = x_i \left(a_i + \sum_{j=1}^{n} b_{ij} x_j \right), \qquad 1 \leq i \leq n, \tag{3.1}$$

constitutes the simplest nonlinear model of interacting multispecies population dynamics. In (3.1), the intrinsic growth rates a_i and the interaction rates b_{ij} are assumed, in the simplest case, to be constants whose signs indicate whether the model represents prey-predator, competition, mutualism, or some mixture of these population dynamics types. The solution

$$x(t; x) = \{x_i(t; x)\}$$

of (3.1) emanating from the point $x(0; x) = x$ in the positive cone

$$R_+^n = \{x \in R^n : x_i > 0, 1 \leq i \leq n\}$$

represents the population densities of the species ensemble t time units after the levels x are attained. Elementary ordinary differential equations theory indicates that, for $x \in R_+^n$, the solution of (3.1)

$$x(t) = x(t; x) \in R_+^n, \qquad t \in [0, T),$$

for some $T > 0$, where either

$$T = \infty \tag{3.2}$$

or, for some j, $1 \leq j \leq n$,

$$\lim_{t \to T} x_j(t) = \infty. \tag{3.3}$$

Further, the qualitative theory of ordinary differential equations addresses the question of the behavior of the solution as t approaches T; for example, boundedness conditions rule out the possibility (3.3) and consequently imply global existence (3.2) of the solution. Usually, for (3.1), the existence of an equilibrium $\bar{x} = \{\bar{x}_i\} \in R_+^n$ is assumed; this requires

$$a_i + \sum_{j=1}^{n} b_{ij} \bar{x}_j = 0, \qquad 1 \leq i \leq n. \tag{3.4}$$

With $a = \{a_i\}$ and $B = \{b_{ij}\}$, (3.4) has the vector form

$$a + B\bar{x} = 0. \tag{3.5}$$

The question of stability of the equilibrium solution $x(t) \equiv \bar{x}$ is, then, fundamental to the qualitative investigation of (3.1). One typical result yields, in this case, that $\bar{x}$ is globally stable in R_+^n if there is a diagonal matrix

$$C = \text{diag}\{c_1, \ldots, c_n\}$$

with $c_i > 0$, $1 \le i \le n$, such that the matrix

$$CB + B^T C \text{ is negative definite} \tag{3.6}$$

(recall that superscript T on a matrix denotes transpose); this result is established (Theorem 5.5 in R_+^n) via the Lyapunov function

$$V(x) = \sum_{i=1}^{n} c_i \left(x_i - \bar{x}_i - \bar{x}_i \ln \frac{x_i}{\bar{x}_i} \right) \tag{3.7}$$

for which

$$\dot{V}(x) = \sum_{i=1}^{n} c_i \left(\frac{x_i - \bar{x}_i}{x_i} \right) \frac{dx_i}{dt} = \frac{1}{2}(x - \bar{x}) \cdot [CB + B^T C](x - \bar{x}) \tag{3.8}$$

(Goh, 1977). The next example illustrates this result.

EXAMPLE 6.2   (Lotka-Volterra Simple Food Chain)   Consider the system

$$\frac{dx_1}{dt} = x_1(a_1 - b_{11}x_1 - b_{12}x_2)$$

$$\frac{dx_2}{dt} = x_2(-a_2 + b_{21}x_1 - b_{22}x_2 - b_{23}x_3) \tag{3.9}$$

$$\frac{dx_3}{dt} = x_3(-a_3 + b_{32}x_2 - b_{33}x_3)$$

where x_1, x_2, and x_3 represent, respectively, the population densities of prey, intermediate predator, and top predator. In this example, the a_i and b_{ij} are positive constants; and $\bar{x}$ exists if

$$a_1 - (b_{11}/b_{21})a_2 - ((b_{11}b_{22} + b_{12}b_{21})/b_{21}b_{32})a_3 > 0. \tag{3.10}$$

Taking

$$B = \begin{bmatrix} -b_{11} & -b_{12} & 0 \\ b_{21} & -b_{22} & -b_{23} \\ 0 & b_{32} & -b_{33} \end{bmatrix}$$

and $C = \text{diag}\{c_1, c_2, c_3\}$ where c_1, c_2, and c_3 are any positive numbers satisfying

$$c_1 b_{12} - c_2 b_{21} = 0 = c_2 b_{23} - c_3 b_{32}, \qquad (3.11)$$

$$CB + B^T C = -2 \begin{bmatrix} c_1 b_{11} & 0 & \\ O & c_2 b_{22} & 0 \\ 0 & 0 & c_3 b_{33} \end{bmatrix}.$$

Thus (3.6) is verified without the need of further assumptions on the interaction matrix B. So whenever (3.10) is satisfied the feasible equilibrium of (3.9) is globally asymptotically stable.

6.3.2 Stochastic Multispecies Models: Nondegenerate Case

To take into account the effects of random environmental fluctuations, stochastic differential equation analogues

$$dX_i = X_i \left(a_i + \sum_{j=1}^{n} b_{ij} X_j \right) dt + \sum_{k=1}^{m} g_{ik}(X)\, dW_k \qquad (3.12)$$

of (3.1) are introduced. In (3.12) the noise intensities g_{ij} are assumed sufficiently smooth, as usual, so local existence and uniqueness of solutions is guaranteed; their specific analytic form coincides with how noise is assumed to affect basic growth and interaction parameters, as in Section 6.1 and 6.2 for single-species models. In this and the next sections the stability characteristics of (3.12) and (3.1) are compared under particular assumptions on the noise intensities. It is perhaps appropriate at this point to issue a warning concerning the applicability of models of the type (3.1) and (3.12). For large numbers of species, parameter estimation problems associated with validation of specific models of these types are considerable. Hence such models have limited use for making concrete predictions for specific systems. Their true value is mainly theoretical in providing possible explanations, hypothesis generators, and standards of comparison for phenomena observed in natural systems.

A natural choice for the functions g_{ij} is the linear homogeneous form

$$g_{ij}(x) = \bar{g}_{ij} x_i, \qquad (3.13)$$

where the $\bar{g}_{ij}$ are constants. Heuristically, this choice arises as a result of replacing each intrinsic growth rate a_i in the deterministic model (3.1) by the sum of an average value and a random fluctuation term

$$a_i + \sum_{j=1}^{m} \bar{g}_{ij} \mathcal{N}_j(t)$$

(here the $\mathcal{N}_j$ are assumed to be independent standard white noise processes); the appropriate interpretation of the resulting model leads to (3.12) with the g_{ij} given by (3.13), analogously to the single population case in Section 2. (Limit theorems in Section 1 indicate precisely how (3.12), (3.13) may arise.) The basic theory of stochastic differential equations, developed in Chapter 3, yields fundamental facts about (3.12) with (3.13). First of all, for $x = \{x_i\} \in R_+^n$, there exists a unique solution

$$X(t) = \{X_i(t)\}$$

of (3.12), (3.13) defined on some random interval $[0, \tau)$ and satisfying $X(0) = x$; the solution $X(t)$ is a Markov diffusion process characterized by the infinitesimal statistics:

$$\lim_{h \to 0^+} \frac{1}{h} E(X(t+h) - X(t)|X(t) = x)$$

$$= \left\{ x_i \left(a_i + \sum_{j=1}^{m} b_{ij} x_j \right) \right\} \qquad \text{(drift coefficient)}$$

$$\lim_{h \to 0} \frac{1}{h} E([X(t+h) - X(t)][X(t+h) - X(t)]^T | X(t) = x)$$

$$= \left\{ \left(\sum_{k=1}^{m} \bar{g}_{ik} \bar{g}_{jk} \right) x_i x_j \right\}. \qquad \text{(diffusion matrix)}$$

It is reasonable to consider solutions of (3.12), (3.13) in some arbitrary bounded region D_0 with smooth boundary ∂D_0 in the feasible region R_+^n; for example D_0 could be chosen so that a solution exiting D_0 would exhibit some component becoming very small or very large—a multispecies model interpretation would be that some species effectively becomes extinct or explodes. The random stopping time τ can be taken as the first exit time from D_0

$$\tau = \inf \{t | X(t) \notin D_0\},$$

and hence may correspond to the time of a significant ecological effect. Suppose now that

$$\det \left(\left\{ \sum_{k=1}^{m} \bar{g}_{ik} \bar{g}_{jk} \right\} \right) \neq 0; \qquad (3.14)$$

then the diffusion matrix

$$\left\{ \sum_{k=1}^{m} g_{ik}(x) g_{jk}(x) \right\} = \left\{ \sum_{k=1}^{m} \bar{g}_{ik} \bar{g}_{jk} x_i x_j \right\}$$

is positive definite for each $x \in R_+^n$, and hence uniformly positive definite on D_0 if

$$\overline{D}_0 = D_0 \cup \partial D_0 \subseteq R_+^n.$$

The Ito operator

$$L = \sum_{i=1}^{n} x_i \left(a_i + \sum_{j=1}^{n} b_{ij} x_j \right) \frac{\partial}{\partial x_i} + \frac{1}{2} \sum_{i,j=1}^{n} \left(\sum_{k=1}^{m} \bar{g}_{ik} \bar{g}_{jk} x_i x_j \right) \frac{\partial^2}{\partial x_i \partial x_j}$$

is then uniformly elliptic on D_0, and so the results of Section 3.5 apply. In particular

$$\tau < \infty \qquad \text{w.p. } 1, \tag{3.15}$$

and the distribution of τ and the distribution of the exit points of X on the boundary ∂D_0 are characterized as solutions of initial-boundary value problems involving L. The biological interpretation of (3.15), therefore, is that if the population density configuration is in D_0 at some time, at least some species effectively becomes extinct or explodes in finite time, and the probabilities of these events can be calculated by solving certain related initial-boundary value problems. In particular, (3.15) implies that stability in the deterministic system is completely destroyed in the stochastic counterpart, although it has been suggested that the mean of τ serves as a measure of relative persistence or stability in this case. Also, under the preceding assumptions, the best possible stability features are stochastic boundedness and the existence of a stable invariant distribution with nowhere zero density in R_+^n; besides (3.14) some size restrictions on the $\bar{g}_{ij}$ are required. Special cases have been given by Barra et al. (1979) and Polansky (1979). To illustrate, the following example is considered.

EXAMPLE 6.3 (Stochastic Food Chain) Consider the system

$$dX_1 = X_1(a_1 - b_{11}X_1 - b_{12}X_2)\, dt + \bar{g}_1 X_1\, dW_1(t)$$

$$dX_2 = X_2(-a_2 + b_{21}X_1 - b_{22}X_2 - b_{23}X_3)\, dt + \bar{g}_2 X_2\, dW_2(t) \tag{3.16}$$

$$dX_3 = X_3(-a_3 + b_{32}X_2 - b_{33}X_3)\, dt + \bar{g}_3 X_3\, dW_3(t).$$

This system represents a stochastic food chain with three species' intrinsic growth rates exhibiting independent random fluctuations with intensities proportional to population densities. Assume the same conditions on the constants a_i, b_{ij}, and c_i as in the deterministic food chain Example 6.2; also take V as in that example, namely,

$$V(x) = \sum_{i=1}^{n} c_i \left[x_i - \bar{x}_i - \bar{x}_i \ln\left(\frac{x_i}{\bar{x}_i}\right) \right].$$

Toward applying Theorem 5.9 with $D = R_+^3$, one calculates

$$LV(x) = \frac{1}{2} \left\{ (x - \bar{x}) \cdot [CB + B^T C](x - \bar{x}) + \sum_{i=1}^{3} c_i \bar{g}_i^2 \bar{x}_i \right\}. \tag{3.17}$$

Letting

$$\delta = \sum_{i=1}^{3} c_i \bar{g}_i^2 \bar{x}_i, \tag{3.18}$$

if the ellipsoid

$$(x - \bar{x}) \cdot [CB + B^T C](x - \bar{x}) + \delta = 0 \tag{3.19}$$

lies entirely in R_+^3, one can take D_{n_0} to be any neighborhood of the ellipsoid with $\overline{D}_{n_0} \subseteq R_+^3$, and the conditions of Theorem 5.9 are met; note that the required nondegeneracy condition follows from (3.13). Therefore, if the ellipsoid (3.19) is contained in R_+^3, the stochastic system (3.16) admits a stable invariant distribution with nowhere zero density in R_+^3. Toward determining a specific condition on the $\bar{g}_i$ that implies this, note that the choice of c_i in Example 6.2 leads to (3.19) written as

$$-2 \sum_{i=1}^{3} c_i b_{ii} (x_i - \bar{x}_i)^2 + \delta = 0; \tag{3.20}$$

for this ellipsoid to lie in R_+^3 it is necessary and sufficient that the points on the semiaxes nearest the bounding coordinate planes $\{x_i = 0\}$, lie in R_+^3, that is,

$$\bar{x}_i - \sqrt{\delta/2c_i b_{ii}} > 0, \qquad i = 1, 2, 3. \tag{3.21}$$

Observe that the explicit size restriction on the $\bar{g}_i$,

$$\sum_{i=1}^{3} c_i \bar{g}_1^2 \bar{x}_i < 2 \min_i \left\{ c_i b_{ii} \bar{x}_i^2 \right\}, \tag{3.22}$$

implies (3.21). Summarizing, under conditions which yield global stability for the corresponding deterministic model, the existence of an invariant distribution for (3.16) is assured if the noise intensity proportionality constants $\bar{g}_i$ satisfy (3.22).

6.3.3 Degenerate Noise Intensity: A Special Case

Hypothesis (3.14) does not hold if, for example,

$$g_{ij}(x) = \bar{g}_{ij}x_i|x - \bar{x}| \tag{3.23}$$

where, for each i and j, $\bar{g}_{ij}$ is a nonzero constant and $x \in R_+^n$. If, in addition, $\bar{x}$ is an equilibrium for the corresponding deterministic model (3.1), then $\bar{x}$ is also an equilibrium for the stochastic model (3.12). Here the solution $X(t) \equiv \bar{x}$ corresponds to an invariant distribution in R_+^n with Dirac density $\delta(x - \bar{x})$, as opposed to the nowhere zero density distribution asserted by the nondegenerate case result mentioned in the last section.

Analogously to the nondegenerate case, (3.23) arises if one replaces each intrinsic growth rate a_i in (3.1) by an average value plus a random fluctuation term

$$a_i + \sum_{j=1}^{m} \bar{g}_{ij}|x - \bar{x}|\mathcal{N}_j$$

where the $\mathcal{N}_j$ are independent standard white noise processes. From the biological point of view, (3.23) runs contrary to the apparently more palatable assumption (3.13), the example of nondegenerate noise discussed in the previous section. Indeed there seems to be no specific biological justification for the existence of such a stochastic equilibrium. Suppose, however, that one is satisfied with the performance of the deterministic model (3.1) in some neighborhood $\tilde{D}$ of the equilibrium $\bar{x}$, but, nevertheless, is also convinced of the utility of the nondegenerate stochastic model (3.12), (3.13) in $R_+^n - \tilde{D}$. (In $R_+^n - \tilde{D}$ some component population may be either large or small and hence possibly more susceptible to random environmental changes such as in weather and food supply.) Then (3.12), (3.23) can be viewed as interpolating between the deterministic and stochastic models, thus providing an alternative to piecing together the models along the boundary of $\tilde{D}$. Note that, nevertheless, under (3.23) $\{\sum_{k=1}^{n} g_{ik}(x)g_{jk}(x)\}$ is positive definite for $x \in D_0 \setminus B_\varepsilon(\bar{x})$, for sufficiently small $\varepsilon > 0$, where D_0 is a neighborhood of $\bar{x}$ and $\overline{D}_0 \subseteq R_+^n$, and $B_\varepsilon(\bar{x}) = \{x \in R^n : |x - \bar{x}| < \varepsilon\}$; so the results concerning first exit time and exit points mentioned in the last section hold for solutions of (3.12), (3.23) with initial points in any annular region A with smooth boundary ∂A and closure $\bar{A} \subseteq D_0 \setminus \{\bar{x}\}$. Together with the stability result described below, these results allow the computation of estimates for probabilities of population extinction and explosion, given a reasonable choice for A.

Stochastic stability, as developed by Has'minskii, and summarized in Chapter 5, is the appropriate analogue of deterministic equilibrium stability in this case.

In particular, the next result can be viewed as a stochastic analogue of Goh's result (given in Section 6.3.1).

THEOREM 6.5   Assume (3.5) and (3.23). Suppose also there is a positive diagonal matrix $C = \text{diag}\{c_1, \ldots, c_n\}$ such that the matrix

$$CB + B^T C + \text{tr}(C\bar{G}\bar{G}^T \bar{X})I \text{ is negative definite}, \tag{3.24}$$

where $\bar{G} = \{\bar{g}_{ij}\}$, $\bar{X} = \text{diag}\{\bar{x}_1, \ldots, \bar{x}_n\}$, and I is the $n \times n$ identity matrix. Then $\bar{x}$ is stochastically asymptotically stable in the large.

Proof: With

$$V(x) = \sum_{i=1}^{n} c_i \left[x_i - \bar{x}_i - \bar{x}_i \ln\left(\frac{x_i}{\bar{x}_i}\right) \right],$$

as in Goh's result, one obtains

$$LV(x) = \frac{1}{2} \left\{ (x - \bar{x}) \cdot [CB + B^T C](x - \bar{x}) \right.$$

$$\left. + \sum_{i=1}^{n} \sum_{j=1}^{m} g_{ij}^2(x) c_i \left(\frac{\bar{x}_i}{x_i^2}\right) \right\}.$$

By using (3.23), the last term in the bracket can be written

$$\left(\sum_{i=1}^{n} \sum_{j=1}^{m} \bar{g}_{ij}^2 c_i \bar{x}_i \right) |x - \bar{x}|^2 = (x - \bar{x}) \cdot \text{tr}(C\bar{G}\bar{G}^T \bar{X})I(x - \bar{x}).$$

Therefore (3.24) implies LV is negative definite as a function of $x - \bar{x}$ in R_+^n, and so the result follows from Theorem 5.7.   □

EXAMPLE 6.4   (Stochastic Lotka-Volterra Food Chain)   Consider the stochastic system

$$dX_1 = X_1([a_1 - b_{11}X_1 - b_{12}X_2]\,dt + \bar{g}_1|X - \bar{x}|\,dW_1(t))$$

$$dX_2 = X_2([-a_2 + b_{21}X_1 - b_{22}X_2 + b_{23}X_3]\,dt$$

$$+ \bar{g}_2|X - \bar{x}|\,dW_2(t)) \tag{3.25}$$

$$dX_3 = X_3([-a_3 + b_{32}X_2 - b_{33}X_3]\,dt + \bar{g}_3|X - \bar{x}|\,dW_3(t)).$$

This system represents a stochastic food chain with the three species' intrinsic growth rates exhibiting independent random fluctuations with intensities proportional to the product of species density and distance from equilibrium of the species ensemble. Assume again the same conditions on the a_i, b_{ij}, and the c_i as employed in Example 6.2, namely, (3.10) and (3.11). Then one can compute

$$CB + B^T C + \text{tr}(C\bar{G}\bar{G}^T \bar{X})I$$
$$= -\operatorname{diag}\{2c_1 b_{11} - \delta, 2c_2 b_{22} - \delta, 2c_3 b_{33} - \delta\}$$

where

$$\delta = \sum_{i=1}^{3} c_i \bar{g}_i^2 \bar{x}_i.$$

Therefore, if

$$\delta < 2 \min_i c_i b_{ii} \tag{3.26}$$

(3.24) holds and Theorem 6.5 can be applied. For the particular choice

$$c_1 = \frac{b_{21}}{b_{12}}, \qquad c_2 = 1, \qquad c_3 = \frac{b_{23}}{b_{32}}$$

(3.26) has the specific form

$$\frac{b_{21}}{b_{12}} \bar{g}_1^2 \bar{x}_1 + \bar{g}_2^2 \bar{x}_2 + \frac{b_{23}}{b_{32}} \bar{g}_3^2 \bar{x}_3 < 2 \min \left\{ \frac{b_{21}}{b_{12}} b_{11}, b_{22}, \frac{b_{23}}{b_{32}} b_{33} \right\}. \tag{3.27}$$

Thus (3.10) and (3.27) yield global stochastic asymptotic stability of the equilibrium solution $X = \bar{x}$ of (3.25).

EXERCISES

6.1 Verify (1.13).

6.2 By using the boundary classification scheme given in Section 5.4, verify the statements about the nature of the boundaries at 0 and ∞ for both the Ito and Stratonovich solutions of the Eq. (2.1) made at the beginning of this section, that is,

(a) Show that 0 and ∞ are unattainable for both solutions.

(b) Show that conditions (2.4) and (2.5) determine which boundary point is attracting in each case. [Hint: See Example 5.5.]

6.3 (a) Verify that solutions $x(t)$ of the deterministic logistic equation

$$\frac{dx}{dt} = rx \left(1 - \frac{x}{K}\right)$$

with $0 < x(0) < K$ satisfy

(i) $0 < x(t) < K$, all $t > 0$.

(ii) $\lim_{t \to \infty} x(t) = K$.

(b)   Show that K is an unattainable attracting boundary for both Ito and Stratonovich solutions of (2.7).

(c)   Prove that for any number $M > 0$,

$$X(0) > 0 \Longrightarrow P(\sup X(t) > M) > 0,$$

for either type of solution of either (2.8) or (2.9). [Hint: See Example 5.6.]

6.4   Find sufficient conditions for the existence of an invariant distribution with nowhere zero density in R_+^2 for the Ito-interpreted stochastic competition system

$$dX_1 = X_1(a_1 - b_{11}X_1 - b_{12}X_2)\,dt + \bar{g}_1 X_1\,dW_1(t)$$

$$dX_2 = X_2(a_2 - b_{21}X_1 - b_{22}X_2)\,dt + \bar{g}_2 X_2\,dW_2(t)$$

where W_1 and W_2 are independent Wiener processes. [Hint: See Example 6.3.]

6.5   Find sufficient conditions for global stochastic asymptotic stability of the feasible equilibrium $\bar{x} = (\bar{x}_1, \bar{x}_2)$ for the competition system

$$dX_1 = X_1(a_1 - b_{11}X_1 - b_{12}X_2)\,dt + \bar{g}_1 X_1 |X - \bar{x}|\,dW_1(t)$$

$$dX_2 = X_2(a_2 - b_{21}X_1 - b_{22}X_2)\,dt + \bar{g}_2 X_2 |X - \bar{x}|\,dW_2(t)$$

where W_1 and W_2 are independent Wiener processes. [Hint: See Example 6.4.]

7

Quantitative Theory of
Stochastic Differential Equations:
Sample Path Approximations

7.0 INTRODUCTION

Once a differential equation model for some physical phenomenon is formulated, one of course wants to obtain the exact solution. Even for ordinary differential equations, this can be accomplished generally only for the simplest (linear) equations. The development in Chapter 4 indicates additional restrictions that accompany the stochastic equations case. In lieu of such a precise representation for the solution, qualitative theory methods (Chapter 5) can be used to establish robust solution properties, such as stability. A basic technique here is to estimate the equation by one with known properties (the method of differential inequalities). On the other hand, quantitative or numerical methods supply approximations of the solutions themselves. These techniques provide specific solution information when, for example, initial data are known. Implementing numerical schemes using several initial data may suggest certain general solution properties that can be further investigated by qualitative methods as well.

A *numerical method* for solving the ordinary initial value problem

$$\frac{dx}{dt} = f(t, x)$$
$$x(t_0) = x_0$$

(0.1)

on some interval $[t_0, T]$ is a scheme for generating a sequence of numbers $\{\bar{x}_n\}$ which approximate the solution values $\{x(t_n)\}$ at prescribed points

t_n in $[t_0, T]$. For the stochastic initial value problem

$$dX = f(t, X)\, dt + g(t, X)\, dW$$
$$X(t_0) = X_0$$

(0.2)

the analogous notion is to obtain sequences which approximate the sample paths of the solution $X(t)$: samples of discrete parameter random processes $\bar{x}_n$ are generated as approximates to solution process sample values $X(t_n)$. Numerical schemes for (0.1) and (0.2) are usually described by deterministic and stochastic difference equations respectively. [See (6.1.24) for an example.] The numerical approximate values are generated iteratively from the difference equations associated with the increments

$$\Delta t_n = t_n - t_{n-1}$$

corresponding to the chosen interval partition

$$t_0 < t_1 < \cdots < t_n < \cdots < t_N = T.$$

For the stochastic case the standard Wiener increments

$$\Delta W_n = W(t_n) - W(t_{n-1})$$

are required as well; these are obtained as sample values of mean zero, variance Δt_n normal random variables. Such stochastic schemes are computer implementable, therefore, provided a random number generator is at hand. The difference equations describing these schemes are suggested by the integral equation for the true increment

$$X(t_n) - X(t_{n-1}) = \int_{t_{n-1}}^{t_n} f(t, X(t))\, dt + \int_{t_{n-1}}^{t_n} g(t, X(t))\, dW(t)$$

(0.3)

of the solution, in particular, by numerical quadrature methods for the stochastic integrals appearing in (0.3). The simplest such method involves the integral approximates, given in Chapter 2, in the definition of the stochastic integral.

The most important questions discussed in the theory of quantitative methods are those concerning convergence, local and global error estimates, and numerical stability. [Convergence of the stochastic numerical scheme represented by (6.1.24), for example, is given by Theorem 6.3.] Numerical stability means that calculation errors do not propagate in an unbounded manner. This chapter focuses on the local or one-step truncation error, specifically, on the order of magnitude of the local error. That is, what is discussed here concerns mainly how the difference between the exact solution value $X(t_n)$ and the numerical approximate value $\bar{X}_n$, as measured by the quadratic mean squared expression

$$E|X(t_n) - \bar{X}_n|^2, \qquad (0.4)$$

is related to the corresponding increment Δt_n, given that $X(t_{n-1}) = \bar{X}_{n-1}$. With (0.4) the error order of magnitudes for the stochastic numerical schemes are the same as for the analogous deterministic ones. However, since (0.4) is the square of the quadratic mean, these results indicate that the orders of magnitude for the stochastic schemes are really half what one obtains from the corresponding deterministic methods. This is not surprising when one recalls that the Wiener increment ΔW_n behaves like $(\Delta t_n)^{1/2}$ in the quadratic mean sense. Specifically, stochastic methods corresponding to the Euler, Heun (or modified Euler), and the higher-order Runge-Kutta methods for deterministic equations are discussed in Sections 7.1, 7.2 and 7.3, respectively, of this chapter. The nth step of these methods is characterized by a polynomial in Δt_n and ΔW_n. A main point that is emphasized here is that, unlike the deterministic case, one cannot generally achieve a higher order than $O((\Delta t_n)^3)$ one-step error order of magnitude by using higher-degree polynomials (corresponding to more function evaluations). In Section 7.4 it is shown that for systems (or vector equations)

$$dX = f(t, X)\, dt + \sum_{j=1}^{m} g_j(t, X)\, dW_j$$

even this order of magnitude for the error can be attained only if the vector-valued functions g_j satisfy a certain symmetry condition involving their gradients. Section 7.5 contains a description of how this difficulty may be overcome by adding to the numerical scheme terms which correspond to iterated stochastic integrals. An important problem, not dealt with in this chapter, that is associated with implementing numerical integration schemes for stochastic equations, is sample error; Klauder and Petersen (1985), for example, give a discussion of sample error for a specific Runge-Kutta type scheme. Finally, it should be noted that the sample path simulation approach given in this chapter can be viewed as an alternative to approximating the distribution of $X(t)$ by numerically integrating the backward or forward PDEs [(1.8.8) or (1.8.9)] for the transition probability density of $X(t)$.

The development in this chapter follows primarily the papers of Mil'shtein (1974), Rümelin (1982), and Wagner and Platen (1978). To begin, the next section presents a full account, concluding with a discussion of numerical stability, of the Euler method for stochastic differential equations.

7.1   EULER'S METHOD

The most basic method for numerically solving ordinary differential equations

$$\frac{dx}{dt} = f(t, x) \tag{1.1}$$

is Euler's method. Consider a partition

$$t_0 < t_1 < \cdots < t_N = T$$

of the finite interval $[t_0, T]$. The values $x(t_n)$ of the solution $x(t) = x(t; x_0)$ of the initial value problem (1.1),

$$x(t_0) = x_0 \tag{1.2}$$

defined on the interval $[t_0, T]$, are approximated by the values $\{\bar{x}_n\}$ given by

$$\begin{aligned} \bar{x}_0 &= x_0 \\ \bar{x}_n &= \bar{x}_{n-1} + f(t_{n-1}, \bar{x}_{n-1})\, \Delta t_n, \end{aligned} \tag{1.3}$$

$n = 1, \ldots, N$, where $\Delta t_n = t_n - t_{n-1}$. The (global) truncation error associated with Euler's method at t_n is given by

$$|x(t_n) - \bar{x}_n|; \tag{1.4}$$

if one assumes $\bar{x}_{n-1} = x(t_{n-1})$, then (1.4) represents the local or one-step error of the approximation at t_n. A continuous approximation can be constructed by taking the piecewise linear interpolation

$$\begin{aligned} \bar{x}(t) &= \bar{x}_{n-1} + (t - t_{n-1}) f(t_{n-1}, \bar{x}_{n-1}), \qquad t_{n-1} < t \leq t_n, \\ \bar{x}(t_0) &= x_0 \end{aligned} \tag{1.5}$$

of (1.3). It is convenient to work with equally spaced partitions with common increment designated by

$$h = \Delta t_n = t_n - t_{n-1} = 1/N.$$

The following well-known result summarizes the basic theory for Euler's method.

THEOREM 7.1   Assume that f is Lipschitz continuous in both variables: there is a constant K such that for all s, $t \in [t_0, T]$, x, $y \in R$,

$$|f(s,x) - f(t,x)| \leq K|s - t|$$

and   (1.6)

$$|f(t,x) - f(t,y)| \leq K|x - y|.$$

Then $\bar{x}(t)$ given by (1.5) converges uniformly to the true solution $x(t)$ of (1.1), (1.2) on $[t_0, T]$, and furthermore, the local and global errors have orders of magnitude $O(h^2)$ and $O(h)$ respectively as $h \to 0$.

For the stochastic initial value problem

$$dX = f(t, X)\, dt + g(t, X)\, dW \tag{1.7}$$

$$X(t_0) = X_0, \tag{1.8}$$

the analogue of Euler's method is the Euler-Maruyama scheme,

$$\bar{X}_0 = X_0$$
$$\bar{X}_n = \bar{X}_{n-1} + f(t_{n-1}, \bar{X}_{n-1})\,\Delta t_n + g(t_{n-1}, \bar{X}_{n-1})\,\Delta W_n \tag{1.9}$$
$$n = 1, \ldots, N,$$

where ΔW_n denotes the Wiener increment $W(t_n) - W(t_{n-1})$. If $X(t)$ denotes the true solution of (1.7), (1.8), the (local) error associated with scheme (1.9) at t_n can be taken as the conditional quadratic mean squared difference

$$E(|X(t_n) - \bar{X}_n|^2 | X(t_{n-1}) = \bar{X}_{n-1} = \bar{x}_{n-1}) \tag{1.10}$$

where $\bar{x}_{n-1}$ is an arbitrary real vlaue. This choice for measuring the difference between the true solution and the approximate is appropriate given the fundamental role played by the mean square norm in stochastic problems, the ease of obtaining the desired error estimates it affords, and that the results it leads to coincide with the corresponding deterministic ones. Chebyshev's inequality verifies that error order of magnitude results obtained for (1.10) imply similar results for the error in probability

$$P[|X(t_n) - \bar{X}_n| > \varepsilon | X(t_{n-1}) = \bar{X}_{n-1} = \bar{x}_{n-1}], \qquad \varepsilon > 0.$$

Conditioning on equality of exact solution value and Euler approximate at the initial value t_0 gives the global error: for any real value $\bar{x}_0$

$$E(|X(t_n) - \bar{X}_n|^2 | X_0 = \bar{X}_0 = \bar{x}_0). \tag{1.11}$$

A corresponding continuous parameter process to (1.9) is given by

$$\bar{X}(t) = \bar{X}_{n-1} + (t - t_{n-1})f(t_{n-1}, \bar{X}_{n-1})$$
$$+ (W(t) - W(t_{n-1}))g(t_{n-1}, \bar{X}_{n-1}), \qquad t_{n-1} < t \leq t_n; \ (1.12)$$

$\bar{X}(t)$ solves the initial value problems

$$d\bar{X}(t) = f(t_{n-1}, \bar{X}_{n-1})\, dt + g(t_{n-1}, \bar{X}_{n-1})\, dW_n(t), \qquad t_{n-1} < t \leq t_n,$$
$$\bar{X}(t_{n-1}) = \bar{X}_{n-1},$$

where W_n is a Wiener process on $[t_{n-1}, \infty)$ so that $W_n(t_{n-1}) = 0$. The next result is the stochastic version of Theorem 7.1. Note that, in addition to the Lipschitz and growth conditions on f and g in the second variable which guarantee existence and uniqueness of the solution of the stochastic initial value problem (1.7), (1.8), a modulus of continuity condition in the t variable is required to obtain similar order of magnitude results for the error as in Theorem 7.1. Convergence here was first established by Maruyama. Once again, it is assumed that the partition of $[t_0, T]$ in (1.9) is equally spaced with common increment $\Delta t_n = h$.

THEOREM 7.2   Suppose the functions f and g satisfy uniform growth and Lipschitz conditions in the second variable, and are Hölder continuous of order $1/2$ in the first variable: there is a constant K such that for all $s, t \in [t_0, T]$, $x, y \in R$,

$$|f(t, x) - f(t, y)| + |g(t, x) - g(t, y)| \leq K|x - y| \tag{1.13a}$$

$$|f(t, x)|^2 + |g(t, x)|^2 \leq K^2(1 + |x|^2) \tag{1.13b}$$

$$|f(s, x) - f(t, x)| + |g(s, x) - g(t, x)| \leq K|s - t|^{1/2}. \tag{1.13c}$$

Then $\bar{X}(t) \to X(t)$ in the mean square sense uniformly on $[t_0, T]$, and the local and global errors, (1.10) and (1.11), have orders of magnitude $O(h^2)$ and $O(h)$, respectively, all as $h \to 0$.

Proof: It suffices to establish the claimed orders of magnitude for the errors. To begin, denote $E|X(t_n) - \bar{X}_n|^2$ by $\mathcal{E}_n$, apply Ito's formula, and take expected value to obtain

$$\mathcal{E}_n = \mathcal{E}_{n-1} + \int_{t_{n-1}}^{t_n} E\left\{2(X(s) - \bar{X}(s))(f(s, X(s)) - f(t_{n-1}, \bar{X}_{n-1}))\right.$$
$$\left. + (g(s, X(s)) - g(t_{n-1}, \bar{X}_{n-1}))^2\right\} ds$$

$$\leq \mathcal{E}_{n-1} + \int_{t_{n-1}}^{t_n} \left(E|X(s) - \bar{X}(s)|^2 + E|f(s, X(s)) \right.$$
$$\left. - f(t_{n-1}, \bar{X}_{n-1})|^2 + E|g(s, X(s)) - g(t_{n-1}, \bar{X}_{n-1})|^2 \right) ds.$$

(1.14)

Invoking the Lipschitz (1.13a) and Hölder (1.13c) conditions, one has

$$|f(s, X(s)) - f(t_{n-1}, \bar{X}_{n-1})|^2 \leq 4 \left\{ |f(s, X(s)) - f(s, X(t_{n-1}))|^2 \right.$$
$$+ |f(s, X(t_{n-1})) - f(t_{n-1}, X(t_{n-1}))|^2$$
$$+ |f(t_{n-1}, X(t_{n-1})) - f(t_{n-1}, \bar{X}_{n-1})|^2$$
$$\leq 4K^2 \left\{ |X(s) - X(t_{n-1})|^2 + s - t_{n-1} + |X(t_{n-1}) - \bar{X}_{n-1}|^2 \right\},$$

(1.15)

and the same for g. Furthermore, from Theorem 3.8, there is a constant K_1 depending only on K, t_0, T, and X_0, such that

$$E|X(s) - X(t_{n-1})|^2 \leq K_1(s - t_{n-1}).$$

(1.16)

Substituting (1.15) and (1.16) into (1.14) leads to

$$\mathcal{E}_n \leq \mathcal{E}_{n-1} + \int_{t_{n-1}}^{t_n} \left(E|X(s) - \bar{X}(s)|^2 \right.$$
$$\left. + 8K^2 \left\{ (K_1 + 1)(s - t_{n-1}) + \mathcal{E}_{n-1} \right\} \right) ds$$
$$\leq \mathcal{E}_{n-1}(1 + 8K^2 h) + 4K^2(K_1 + 1)h^2$$
$$+ \int_{t_{n-1}}^{t_n} E|X(s) - \bar{X}(s)|^2 \, ds.$$

(1.17)

Now, the Bellman-Gronwall inequality (Lemma 3.2) may be applied (with $a(t) = E|X(t) - \bar{X}(t)|^2$, and

$$b(t) \equiv (E|X(t_{n-1}) - X(t_n)|^2)(1 + 8K^2 h) + 4K^2(K_1 + 1)h^2,$$

on $[t_{n-1}, t_n]$) to obtain from (1.17)

$$\mathcal{E}_n \leq \mathcal{E}_{n-1}(1 + 8K^2 h)$$
$$+ \int_{t_{n-1}}^{t_n} e^{t_n - s} [\mathcal{E}_{n-1}(1 + 8K^2 h) + 4K^2(K_1 + 1)h^2] \, ds$$
$$+ 4K^2(K_1 + 1)h^2$$
$$\leq [\mathcal{E}_{n-1}(1 + 8K^2 h) + 4K^2(K_1 + 1)h^2] e^h.$$

(1.18)

For the local error estimate, the assumption $\mathcal{E}_{n-1} = 0$ implies

$$\mathcal{E}_n \leq 4K^2(K_1 + 1)e^{T - t_0} h^2 = O(h^2).$$

For the global error estimate, iterating (1.18) with $\mathcal{E}_0 = 0$ leads to

$$\mathcal{E}_n \le 4K^2(K_1 + 1)h^2 e^h \left(\frac{1 - r^n}{1 - r} \right), \tag{1.19}$$

where $r = r(h) = (1 + 8K^2 h)e^h$. The expression on the right in (1.19) may be arranged

$$4K^2(K_1 + 1)\left[1 - r^{(T - t_0)/h} \right] \left(\frac{he^h}{1 - r} \right) h. \tag{1.20}$$

Now as $h \to 0$,

$$r^{1/h} = (1 + 8K^2 h)^{1/h} e \longrightarrow e^{8K^2 + 1}$$

and, by l'Hôpital's rule,

$$\frac{he^h}{1 - r} = \frac{h}{e^{-h} - (1 + 8K^2 h)} \longrightarrow \frac{-1}{1 + 8K^2}.$$

Therefore, as $h \to 0$, the product multiplying h in (1.20) tends to

$$4K^2(K_1 + 1)\left[\frac{e^{(8K^2 + 1)(T - t_0)} - 1}{8K^2 + 1} \right].$$

This means that (1.19) implies

$$\mathcal{E}_n = O(h). \quad \square$$

An example follows the next section.

Under the conditions of Theorem 7.1, Euler's method (1.3) for ordinary differential equations (1.1) is also numerically stable. Let $\{\bar{x}_n^{(1)}\}$ denote the sequence generated by (1.3) with x_0 replaced by some value $x_0^{(1)}$. *Numerical stability* means: given $\varepsilon > 0$, there is a $\delta > 0$ such that, for all $h > 0$, and positive integers $n \le (T - t_0)/h$

$$\left| x_0 - x_0^{(1)} \right| \le \delta \Longrightarrow \left| \bar{x}_n - \bar{x}_n^{(1)} \right| \le \varepsilon; \tag{1.21}$$

the propagation $\left| \bar{x}_n - \bar{x}_n^{(1)} \right|$ in the numerical scheme (1.3) of an error $\left| x_0 - x_0^{(1)} \right|$ remains bounded. In the absence of numerical stability, the accumulation of roundoff error accompanying computer implementation of a numerical method can lead to difficulties.

The analogous stochastic scheme (1.9) is *numerically stable in the quadratic mean-squared sense* if: Given $\varepsilon > 0$, there is a $\delta > 0$ such that for all $h > 0$, and positive integers $n \le (T - t_0)/h$,

$$\rho_0 \equiv E\left| X_0 - X_0^{(1)} \right|^2 \le \delta \Longrightarrow \rho_n \equiv E\left| \bar{X}_n - \bar{X}_n^{(1)} \right|^2 \le \varepsilon, \tag{1.22}$$

where here $\bar{X}_n^{(1)}$ denotes the sequence defined by (1.9) with X_0 replaced by $X_0^{(1)}$. Verifying that (1.9) is stable under the conditions of Theorem 7.2, follows similarly to the proof of the latter. Indeed, one has

$$\left| \bar{X}_n - \bar{X}_n^{(1)} \right|^2 = \left| \bar{X}_{n-1} - \bar{X}_{n-1}^{(1)} \right|^2$$
$$+ \left[f(t_{n-1}, \bar{X}_{n-1}) - f\left(t_{n-1}, \bar{X}_{n-1}^{(1)}\right) \right]^2 (\Delta t_n)^2$$
$$+ \left[g(t_{n-1}, \bar{X}_{n-1}) - g\left(t_{n-1}, \bar{X}_{n-1}^{(1)}\right) \right]^2 (\Delta W_n)^2$$
$$+ 2 \left[\bar{X}_{n-1} - \bar{X}_{n-1}^{(1)} \right] \left[f(t_{n-1}, \bar{X}_{n-1}) - f\left(t_{n-1}, \bar{X}_{n-1}^{(1)}\right) \right] \Delta t_n$$
$$+ 2 \left[\bar{X}_{n-1} - \bar{X}_{n-1}^{(1)} \right] \left[g(t_{n-1}, \bar{X}_{n-1}) - g\left(t_{n-1}, \bar{X}_{n-1}^{(1)}\right) \right] \Delta W_n$$
$$+ 2 \left[f(t_{n-1}, \bar{X}_{n-1}) - f\left(t_{n-1}, \bar{X}_{n-1}^{(1)}\right) \right] \left[g(t_{n-1}, \bar{X}_{n-1}) \right.$$
$$\left. - g\left(t_{n-1}, \bar{X}_{n-1}^{(1)}\right) \right] \Delta t_n \, \Delta W_n$$
$$\leq \left| \bar{X}_{n-1} - \bar{X}_{n-1}^{(1)} \right|^2 \left[1 + K^2(\Delta t_n)^2 + K^2(\Delta W_n)^2 + (1 + K^2)\Delta t_n \right]$$
$$+ 2 \left[\bar{X}_{n-1} - \bar{X}_{n-1}^{(1)} \right] \left[g(t_{n-1}, \bar{X}_{n-1}) - g(t_{n-1}, \bar{X}_{n-1}^{(1)}) \right] \Delta W_n$$
$$+ 2 \left[f(t_{n-1}, \bar{X}_{n-1}) - f\left(t_{n-1}, \bar{X}_{n-1}^{(1)}\right) \right] \left[g(t_{n-1}, \bar{X}_{n-1}) \right.$$
$$\left. - g\left(t_{n-1}, \bar{X}_{n-1}^{(1)}\right) \right] \Delta t_n \, \Delta W_n \tag{1.23}$$

In obtaining the estimate (1.23), the Lipschitz condition has been used and the Cauchy-Schwarz inequality has been invoked. The factors multiplying ΔW_n and $(\Delta W_n)^2$ in (1.23) are independent of ΔW_n and $(\Delta W_n)^2$ respectively. Since $E\,\Delta W_n = 0$ and $E(\Delta W_n)^2 = \Delta t_n = h$, taking expected value in (1.23) leads to the inequality for the propagated error

$$\rho_n \leq \rho_{n-1}(1 + (2K^2 + 1)h + K^2 h^2). \tag{1.24}$$

Iterating (1.24) yields

$$\rho_n \leq \rho_0 (1 + (2K^2 + 1)h + K^2 h^2)^n$$
$$\leq \rho_0 (1 + (2K^2 + 1)h + K^2 h^2)^{(1/h)(T - t_0)}. \tag{1.25}$$

Since $(1 + (2K^2 + 1)h + K^2 h^2)^{1/h} \uparrow e^{2K^2 + 1}$ as $h \downarrow 0$, (1.25) implies that, given $\varepsilon > 0$,

$$\rho_n \leq \varepsilon \qquad \text{if } \rho_0 \leq \delta \equiv \varepsilon e^{(2K^2+1)(t_0-T)};$$

so numerical stability for the stochastic Euler scheme has been established.

7.2 HEUN'S METHOD

The Euler method for numerically solving ordinary differential equations corresponds to the most basic numerical integration scheme—namely, Riemann sums. The trapezoidal rule gives faster convergence than Riemann sums, and so suggests that the analogous numerical solver (for equally spaced partitions with $h = \Delta t_n$)

$$\bar{x}_n = \bar{x}_{n-1} + \frac{f(t_{n-1}, \bar{x}_{n-1}) + f(t_n, \bar{x}_n)}{2} h \qquad (2.1)$$

would yield improved accuracy over Euler's method. An unpleasant feature of (2.1), however, is that the next iterate $\bar{x}_n$ is given implicitly. A remedy for this difficulty is to replace $\bar{x}_n$ in the right side of (2.1) by the Euler approximate based on t_{n-1} and $\bar{x}_{n-1}$:

$$\tilde{x}_n = \bar{x}_{n-1} + f(t_{n-1}, \bar{x}_{n-1})h; \qquad (2.2)$$

the Heun (or modified Euler) scheme

$$\bar{x}_n = \bar{x}_{n-1} + \tfrac{1}{2}\left[f(t_{n-1}, \bar{x}_{n-1}) + f(t_n, \bar{x}_{n-1} + f(t_{n-1}, \bar{x}_{n-1}))\right] h \qquad (2.3)$$

results. The Heun scheme is the simplest of the Runge-Kutta type methods. These methods obtain improved accuracy, in the form of higher error orders of magnitude in the step size h at the expense of more function evaluations per step. A continuous version of (2.3) is defined on $[t_0, T]$ by

$$\bar{x}(t) = \bar{x}_n + \tfrac{1}{2}\left[f(t_{n-1}, \bar{x}_{n-1}) + f(t_n, \tilde{x}_n)\right](t - t_{n-1}),$$
$$t_{n-1} < t \leq t_n \qquad (2.4)$$

where $\tilde{x}_n = \bar{x}_{n-1} + f(t_{n-1}, \bar{x}_{n-1})h$, and $\bar{x}(t_0) = x_0$.

The next theorem summarizes the basic facts for the Heun method.

THEOREM 7.3 Suppose that f has bounded continuous partial derivatives in both variables. Then $\bar{x}(t)$ given by (2.4) converges uniformly to the true solution $x(t)$ of (1.1), (1.2) on $[t_0, T]$, and the local and global errors have orders of magnitude $O(h^3)$ and $O(h^2)$, respectively, as $h \to 0$.

The Heun method yields an order of magnitude improvement in step size over the Euler in the stochastic as well as deterministic case. For the stochastic initial value problem (1.7), (1.8) the analogous Heun scheme is given by

$$\bar{X}_0 = X_0$$

$$\bar{X}_n = \bar{X}_{n-1} + \tfrac{1}{2}\left[f(t_{n-1}, \bar{X}_{n-1}) + f(t_n, \tilde{X}_n)\right]h$$

$$+ \tfrac{1}{2}\left[g(t_{n-1}, \bar{X}_{n-1}) + g(t_n, \tilde{X}_n)\right]\Delta W_n \qquad (2.5)$$

where $\tilde{X}_n = \bar{X}_{n-1} + f(t_{n-1}, \bar{X}_{n-1})h + g(t_{n-1}, \bar{X}_{n-1})\Delta W_n$.

The form of the last term in (2.5) might lead one to believe that convergence will be to the Stratonovich rather than Ito solution of the equation

$$dX = f(t, X)\, dt + g(t, X)\, dW \qquad (2.6)$$

Indeed, this is the case, and the next result gives the stochastic analogue of Theorem 7.3. In what follows, $X(t)$ denotes the Stratonovich solution of (2.6), or equivalently, the Ito solution of

$$dX = \left[f(t, X) + \frac{1}{2}\left(g\frac{\partial g}{\partial x}\right)(t, X)\right]dt + g(t, X)\, dW, \qquad (2.7)$$

which also satisfies the initial condition

$$X(t_0) = X_0. \qquad (2.8)$$

[As in the Euler case in the last section, define the continuous parameter process $\bar{X}(t)$ corresponding to (2.5) by

$$\bar{X}(t) = \bar{X}_{n-1} + \tfrac{1}{2}\left[f(t_{n-1}, \bar{X}_{n-1}) + f(t_n, \tilde{X}_n)\right](t - t_{n-1})$$

$$+ \tfrac{1}{2}\left[g(t_{n-1}, \bar{X}_{n-1}) + g(t_n, \tilde{X}_n)\right](W(t) - W(t_{n-1})), \qquad (2.9)$$

$$t_{n-1} < t \le t_n,$$

$$\bar{X}(t_0) = X_0.]$$

THEOREM 7.4 Suppose the functions f and g have bounded continuous partial derivatives up to the fourth order in both variables. If $\bar{X}$ is defined by (2.9), $\bar{X}(t) \to X(t)$ in the mean square sense uniformly on $[t_0, T]$, and the local and global errors, (1.10) and (1.11), have orders of magnitude $O(h^3)$ and $O(h^2)$, respectively, as the step size $h \to 0$.

Proof: Only the local error statement in Theorem 7.4 will be verified. Indeed, throughout the remainder of this chapter, local or one-step error order of magnitude will be the focus of the discussion. The local error results indicate restrictions on obtaining higher-order accuracy methods which distinguish the stochastic and deterministic cases.

To establish the local error part of Theorem 7.4, an approach due to Mil'shtein [1974] will be followed. This approach enables verification of the stated local error order of magnitude for the Heun scheme in Theorem 7.4 and also facilitates investigation of other schemes for possible higher-order accuracy. One considers numerical schemes extending (1.9) which, at each step, are characterized by higher-degree polynomials in $h = \Delta t_n$ and $W(h)$ (which has the same distribution as ΔW_n). For example, the scheme

$$\bar{X}_n = \bar{X}_{n-1} + c_1(t_{n-1}, \bar{X}_{n-1})h$$
$$+ c_2(t_{n-1}, \bar{X}_{n-1})W(h) + c_3(t_{n-1}, \bar{X}_{n-1})W^2(h), \qquad (2.10)$$

for some choice of the functions c_1, c_2, and c_3, agrees with the Heun method locally up to quadratic mean squared order $O(h^3)$. This will be established below. First, in line with the emphasis here on Ito equations, consider the Heun scheme (2.5) with f replaced by $F = f - \frac{1}{2}g(\partial g/\partial x)$. According to the theorem, this approximation should converge to the Stratonovich solution of

$$dX = F(t, X)\, dt + g(t, X)\, dW \qquad (2.11)$$

which corresponds to the Ito solution of

$$dX = f(t, X)\, dt + g(t, X)\, dW. \qquad (2.12)$$

Expanding (2.5) in this case by Taylor's formula, one obtains

$$\bar{X}_n = \bar{X}_{n-1} + \frac{1}{2}\left[2F(t_{n-1}, \bar{X}_{n-1}) + O(h) + O(|W(h)|)\right]h$$
$$+ \frac{1}{2}\left[2g(t_{n-1}, \bar{X}_{n-1}) + \left(\frac{\partial g}{\partial x}g\right)(t_{n-1}, \bar{X}_{n-1})W(h)\right.$$
$$\left. + O(h) + O(W^2(h))\right]W(h)$$
$$= \bar{X}_{n-1} + F(t_{n-1}, \bar{X}_{n-1})h$$
$$+ g(t_{n-1}, \bar{X}_{n-1})W(h) + \frac{1}{2}\left(\frac{\partial g}{\partial x}g\right)(t_{n-1}, \bar{X}_{n-1})W^2(h)$$
$$+ O(h^2) + O(h|W(h)|) + O(|W(h)|^3). \qquad (2.13)$$

The last three terms in (2.13) have quadratic mean squared orders h^4, h^3, and h^3, respectively; thus (2.13) verifies that the Heun scheme (2.5) with f replaced by $f - \frac{1}{2}g(\partial g/\partial x)$ is equivalent locally up to quadratic mean squared order $O(h^3)$ to (2.10) with the choices $c_1 = F$, $c_2 = g$, and $c_3 = \frac{1}{2}g(\partial g/\partial x)$. It remains to show that this method, namely,

$$\bar{X}_n = \bar{X}_{n-1} + F(t_{n-1}, \bar{X}_{n-1})h$$

$$+ g(t_{n-1}, \bar{X}_{n-1})W(h) + \frac{1}{2}\left[g\frac{\partial g}{\partial x}\right](t_{n-1}, \bar{X}_{n-1})W^2(h) \qquad (2.14)$$

has one step error order of magnitude $O(h^3)$ in the quadratic mean squared sense.

To accomplish this an analogue of Taylor's theorem for expanding the conditional expectation is employed. Suppose $Y(t)$ solves the (vector) stochastic equation (or stochastic system)

$$dY = \tilde{f}(t, Y)\, dt + \tilde{G}(t, Y)\, dW \qquad (2.15)$$

and let L be the associated Ito operator. If $H(t, y)$ is a continuous function with continuous partial derivatives $\partial H/\partial t$, $\partial H/\partial y_i$, and $\partial^2 H/\partial y_i\, \partial y_j$, one has by Ito's formula, for arbitrary $\bar{y}_{n-1}$,

$$E\left(H(t_n, Y(t_n)) \mid Y(t_{n-1}) = \bar{y}_{n-1}\right) - H(t_{n-1}, \bar{y}_{n-1})$$

$$= \int_{t_{n-1}}^{t_n} E\left(LH(t, Y(t)) \mid Y(t_{n-1}) = \bar{y}_{n-1}\right) dt. \qquad (2.16)$$

In particular if the integrand is bounded, the right side of (2.16) is $O(h)$ (in the usual sense), that is, one has

$$E\left(H(t_n, Y(t_n)) \mid Y(t_{n-1}) = \bar{y}_{n-1}\right) = H(t_{n-1}, \bar{y}_{n-1}) + O(h).$$

$$(2.17)$$

Now suppose LH satisfies the conditions cited above on H. (Note that this requires further differentiability assumptions on $\tilde{f}$ and $\tilde{G}$.) Then one can write (2.16) for LH:

$$E\left(LH(t_n, Y(t_n)) \mid Y(t_{n-1}) = \bar{y}_{n-1}\right) - LH(t_{n-1}, \bar{y}_{n-1})$$

$$= \int_{t_{n-1}}^{t_n} E\left(L^2 H(t, Y(t)) \mid Y(t_{n-1}) = \bar{y}_{n-1}\right) dt, \qquad (2.18)$$

where $L^2 H = L(LH)$. Substituting (2.18) into (2.16) and integrating leads to

$$E\left(H(t_n, Y(t_n)) \mid Y(t_{n-1}) = \bar{y}_{n-1}\right)$$
$$= H(t_{n-1}, \bar{y}_{n-1}) + LH(t_{n-1}, \bar{y}_{n-1})h$$
$$+ \int_{t_{n-1}}^{t_n} \int_{t_{n-1}}^{t} E\left(L^2 H(s, Y(s)) \mid Y(t_{n-1}) = \bar{y}_{n-1}\right) ds\, dt. \qquad (2.19)$$

The corresponding boundedness assumption on

$$E\left(L^2 H(t, Y(t)) \mid Y(t_{n-1}) = \bar{y}_{n-1}\right)$$

gives, from (2.19),

$$E\left(H(t_n, Y(t_n)) \mid Y(t_{n-1}) = \bar{y}_{n-1}\right)$$
$$= H(t_{n-1}, \bar{y}_{n-1}) + LH(t_{n-1}, \bar{y}_{n-1})h + O(h^2).$$

By induction one extends to

$$E\left(H(t_n, Y(t_n)) \mid Y(t_{n-1}) = \bar{y}_{n-1}\right)$$
$$= H(t_{n-1}, \bar{y}_{n-1}) + LH(t_{n-1}, \bar{y}_{n-1})h + \cdots + \frac{1}{k!} L^k H(t_{n-1}, \bar{y}_{n-1})h^k$$
$$+ O(h^{k+1}) \qquad (2.20)$$

for any positive integer k, provided H, $\tilde{f}$, and $\tilde{G}$ are sufficiently smooth. Equation (2.20) is the Taylor-type result to be used.

To obtain the local error order of magnitude for the Heun scheme, one applies (2.20) with the following choices of Y and H. One takes

$$Y(t) = \begin{bmatrix} X(t) \\ W(t) \end{bmatrix}$$

where $X(t)$ is the solution of (2.12) on $[t_{n-1}, t_n]$ and $W(t)$ is the Wiener process; that is, Y solves the trivial system (2.15) where

$$\tilde{f} = \begin{bmatrix} f \\ 0 \end{bmatrix} \quad \text{and} \quad \tilde{G} = \begin{bmatrix} g \\ 1 \end{bmatrix}.$$

Choose $H = \Phi^2$ where

$$\Phi(t, y) = \Phi(t, x, w)$$
$$= x - \bar{x}_{n-1} - c_1(t - t_{n-1}) - c_2(w - \bar{w}_{n-1})$$
$$- c_3(w - \bar{w}_{n-1})^2 \qquad (2.21)$$

so that, with $\bar{y}_{n-1} = (\bar{x}_{n-1}, \bar{w}_{n-1})^T$, $\bar{x}_{n-1}$ and $\bar{w}_{n-1}$ arbitrary,

$$E\left(H(t_n, Y(t_n)) \mid Y(t_{n-1}) = \bar{y}_{n-1}\right)$$
$$= E\left(|X(t_n) - \bar{X}_n|^2 \mid X(t_{n-1}) = \bar{X}_{n-1} = \bar{x}_{n-1}, W(t_{n-1}) = \bar{w}_{n-1}\right),$$

$$(2.22)$$

the local error in the quadratic mean squared sense. It follows that

$$H(t_{n-1}, \bar{y}_{n-1}) = H(t_{n-1}, \bar{x}_{n-1}, \bar{w}_{n-1}) = 0,$$

and, from (2.20), that

$$E\left(|X(t_n) - \bar{X}_n|^2 \mid X(t_{n-1}) = \bar{X}_{n-1} = \bar{x}_{n-1}, W(t_{n-1}) = \bar{w}_{n-1}\right)$$
$$= LH(t_{n-1}, \bar{x}_{n-1}, \bar{w}_{n-1})h + \cdots$$
$$+ \frac{1}{k!} L^k H(t_{n-1}, \bar{x}_{n-1}, \bar{w}_{n-1})h^k + O(h^{k+1}) \qquad (2.23)$$

for any positive integer k if f and g are sufficiently smooth. Here L is the Ito operator associated with the trivial system

$$dX = f(t, X)\, dt + g(t, X)\, dW$$
$$dW = dW, \qquad\qquad\qquad\qquad\qquad\qquad (2.24)$$

namely,

$$L = \frac{\partial}{\partial t} + f\frac{\partial}{\partial x} + \frac{1}{2}g^2\frac{\partial^2}{\partial x^2} + g\frac{\partial^2}{\partial x \partial w} + \frac{1}{2}\frac{\partial^2}{\partial w^2}. \qquad (2.25)$$

So to verify $O(h^{k+1})$ order of magnitude for the local error it suffices to show that

$$LH, \cdots, L^k H$$

all vanish at $(t_{n-1}, \bar{x}_{n-1}, \bar{w}_{n-1})$. For the $k = 2$ case the right side of (2.23) is

$$LH(t_{n-1}, \bar{x}_{n-1}, \bar{w}_{n-1})h + \tfrac{1}{2}L^2 H(t_{n-1}, \bar{x}_{n-1}, \bar{w}_{n-1})h^2 + O(h^3),$$

$$(2.26)$$

with H given by (2.21) provided f and g have continuous fourth-order partial derivatives. To establish the local error order of magnitude $O(h^3)$ it suffices to show

$$LH(t_{n-1}, \bar{x}_{n-1}, \bar{w}_{n-1}) = 0 \text{ and } L^2 H(t_{n-1}, \bar{x}_{n-1}, \bar{w}_{n-1}) = 0. \qquad (2.27)$$

Since $H = \Phi^2$, one has

$$LH = 2\Phi L\Phi + (S\Phi)^2 \tag{2.28a}$$

and

$$L^2 H = 2(L\Phi)^2 + 2\Phi L^2\Phi + 2S\Phi(SL\Phi + LS\Phi) + (S^2\Phi)^2 \tag{2.28b}$$

where

$$S\Phi(t, x, w) = \left[g \frac{\partial \Phi}{\partial x} + \frac{\partial \Phi}{\partial w} \right] (t, x, w).$$

Note that, from (2.28), condition (2.27) holds if

$$\begin{aligned} S\Phi(t_{n-1}, \bar{x}_{n-1}, \bar{w}_{n-1}) &= 0, \\ L\Phi(t_{n-1}, \bar{x}_{n-1}, \bar{w}_{n-1}) &= 0, \\ S^2\Phi(t_{n-1}, \bar{x}_{n-1}, \bar{w}_{n-1}) &= 0. \end{aligned} \tag{2.29}$$

Here

$$\begin{aligned} \Phi(t, x, w) = {}& x - \bar{x}_{n-1} - c_1(t - t_{n-1}) - c_2(w - \bar{w}_{n-1}) \\ & - c_3(w - \bar{w}_{n-1})^2, \end{aligned}$$

so one obtains

$$\begin{aligned} S\Phi(t, x, w) &= g(t, x) - c_2 - 2c_3(w - \bar{w}_{n-1}), \\ L\Phi(t, x, w) &= -c_1 + f(t, x) - c_3, \\ S^2\Phi(t, x, w) &= S(S\Phi(t, x, w)) = g(t, x) \frac{\partial g(t, x)}{\partial x} - 2c_3. \end{aligned} \tag{2.30}$$

From (2.30) it follows that the choices

$$\begin{aligned} c_1 &= \left[f - \tfrac{1}{2} g \frac{\partial g}{\partial x} \right] (t_{n-1}, \bar{x}_{n-1}) \\ c_2 &= g(t_{n-1}, \bar{x}_{n-1}) \\ c_3 &= \frac{1}{2} \left[g \frac{\partial g}{\partial x} \right] (t_{n-1}, \bar{x}_{n-1}) \end{aligned}$$

in the equivalent Heun scheme (2.14), then, are precisely the ones needed to insure (2.29), and hence (2.27). This completes the proof. $\square$

EXAMPLE 7.1   Consider the stochastic equation

$$dX = X\, dW. \tag{2.31}$$

The exact solution of (2.31) is

$$X(t) = X_0 \exp\left\{-\tfrac{1}{2}t + W(t)\right\}. \tag{2.32}$$

The Euler-Maruyama method (1.9) takes the form

$$\begin{aligned} \bar{X}_0 &= X_0 \\ \bar{X}_n &= \bar{X}_{n-1} + \bar{X}_{n-1}\Delta W_n, \end{aligned} \tag{2.33}$$

while the equivalent Heun scheme (2.14) is given by

$$\begin{aligned} \bar{X}_0 &= X_0 \\ \bar{X}_n &= \bar{X}_{n-1} - \tfrac{1}{2}\bar{X}_{n-1}h + \bar{X}_{n-1}\Delta W_n + \tfrac{1}{2}\bar{X}_{n-1}(\Delta W_n)^2. \end{aligned} \tag{2.34}$$

In particular, in one step one has

$$\bar{X}_1 = \bar{X}_0[1 + W(h)] \tag{2.35}$$

and

$$\bar{X}_1 = \bar{X}_0\left[1 - \tfrac{1}{2}h + W(h) + \tfrac{1}{2}W^2(h)\right] \tag{2.36}$$

for (2.33) and (2.24) respectively. The comparable exact solution value

$$X(h) = X_0 \exp\left\{-\tfrac{1}{2}h + W(h)\right\}$$

expanded, via Taylor, yields

$$\begin{aligned} X(h) &= X_0\left[1 - \tfrac{1}{2}h + W(h) + \tfrac{1}{2}\left(-\tfrac{1}{2}h + W(h)\right)^2 \right. \\ &\quad \left. + O((h + |W(h)|)^3)\right] \\ &= X_0\left[1 - \tfrac{1}{2}h + W(h) + \tfrac{1}{2}W^2(h) \right. \\ &\quad \left. + O(h^2) + O(h|W(h)|) + O(|W(h)|^3)\right]. \end{aligned} \tag{2.37}$$

With $\bar{X}_1$ given by (2.35), one obtains the one-step error

$$\begin{aligned} E\left(|X(h) - \bar{X}_1|^2 \mid X_0 = \bar{X}_0 = \bar{x}_0\right) \\ = \bar{x}_0^2 E\left[-\tfrac{1}{2}h + \tfrac{1}{2}W^2(h) + O(h^2) + O(h|W(h)|) + O(|W(h)|^3)\right]^2 \\ = O(h^2), \end{aligned}$$

and for $\bar{X}_1$ given by (2.36), the order of magnitude estimate for the one-step error is

$$
E\left(|X(h) - \bar{X}_1|^2 \mid X_0 = \bar{X}_0 = \bar{x}_0\right)
$$
$$
= \bar{x}_0^2 E\left[O(h^2) + O(h|W(h)|) + O(|W(h)|^3)\right]^2
$$
$$
= O(h^3).
$$

7.3   HIGHER-ORDER RUNGE-KUTTA-TYPE SCHEMES

The Heun (or modified Euler) method is a second-order Runge-Kutta-type numerical scheme. As in the case of ordinary differential equations, one might attempt to achieve higher-order accuracy via higher-order Runge-Kutta schemes. The $(m + 1)$th order such scheme has the form (with $h = \Delta t_n$)

$$
\bar{X}_n = \bar{X}_{n-1} + \sum_{i=0}^{m} p_i F_i h + \sum_{i=0}^{m} q_i G_i \, \Delta W_n
$$

where

$$
\bar{X}_0 = X_0
$$
$$
F_0 = f(t_{n-1} + \alpha_0 h, \bar{X}_{n-1}), \qquad G_0 = g(t_{n-1} + \alpha_0 h, \bar{X}_{n-1}),
$$
$$
X_{n-1}^{(1)} = \bar{X}_{n-1} + \beta_{10} F_0 h + \gamma_{10} G_0 \, \Delta W_n,
$$
$$
F_1 = f\left(t_{n-1} + \alpha_1 h, X_{n-1}^{(1)}\right), \qquad G_1 = g\left(t_{n-1} + \alpha_1 h, X_{n-1}^{(1)}\right),
$$
$$
X_{n-1}^{(2)} = \bar{X}_{n-1} + [\beta_{20} F_0 + \beta_{21} F_1]h + [\gamma_{20} G_0 + \gamma_{21} G_1] \, \Delta W_n, \qquad (3.1)
$$
$$
F_2 = f\left(t_{n-1} + \alpha_2 h, X_{n-1}^{(2)}\right), \qquad G_2 = g\left(t_{n-1} + \alpha_2 h, X_{n-1}^{(2)}\right),
$$

. . . .

$$
X_{n-1}^{(m)} = \bar{X}_{n-1} + \sum_{j=0}^{m-1} \beta_{mj} F_j h + \sum_{j=0}^{m-1} \gamma_{mj} G_j \, \Delta W_n,
$$
$$
F_m = f\left(t_{n-1} + \alpha_m h, X_{n-1}^{(m)}\right), \qquad G_m = g\left(t_{n-1} + \alpha_m h, X_{n-1}^{(m)}\right),
$$

and

$$
\sum_{i=0}^{m} p_i = \sum_{i=0}^{m} q_i = 1.
$$

Rümelin (1982) has established the following convergence result for the general Runge-Kutta method (3.1).

THEOREM 7.5 Suppose f, $\partial f/\partial t$, $\partial f/\partial x$, g, $\partial g/\partial t$, $\partial g/\partial x$, $\partial^2 g/\partial t^2$, $\partial^2 g/\partial t\partial x$, $\partial^2 g/\partial x^2$ are bounded. Then the approximation defined by (the corresponding continuous parameter process to) $\bar{X}_n$ given by (3.1) converges uniformly on $[t_0, T]$ in the quadratic mean sense to the (Ito) solution of

$$dX = [f + \lambda g(\partial g/\partial x)](t, X)\,dt + g(t, X)\,dW \tag{3.2}$$

provided

$$\lambda = \sum_{i=1}^{m} q_i \sum_{j=0}^{i-1} \gamma_{ij}, \qquad m \geq 1. \tag{3.3}$$

The proof consists of expanding the F_i and G_i in Taylor series about $(t_{n-1}, \bar{x}_{n-1})$ through terms of order $h = t_n - t_{n-1}$ and ΔW_n and comparing to the Euler approximation. For details, see Rümelin (1982). Notice that if one replaces f by $f - \lambda g(\partial g/\partial x)$ in (3.1), one obtains via Theorem 7.5 convergence to the solution of the Ito equation

$$dX = f(t, X)\,dt + g(t, X)\,dW. \tag{3.4}$$

The value λ is referred to as the correction factor. The Heun scheme discussed in the previous section corresponds to the $m = 1$ case of (3.1) with the choices $\alpha_0 = 0$, $\alpha_1 = \beta_{10} = \gamma_{10} = 1$, $p_0 = p_1 = q_0 = q_1 = 1/2$, and the correction factor is $\lambda = 1/2$. The stochastic analogue of the classical fourth-order Runge-Kutta scheme, on the other hand, comes from the $m = 3$ case of (3.1) with the choices

$$\alpha_0 = \beta_{20} = \beta_{30} = \beta_{31} = \gamma_{20} = \gamma_{30} = \gamma_{31} = 0,$$

$$\alpha_1 = \alpha_2 = \beta_{10} = \beta_{21} = \gamma_{10} = \gamma_{21} = 1/2, \quad \alpha_3 = \beta_{32} = \gamma_{32} = 1,$$

$$p_0 = p_3 = q_0 = q_3 = 1/6, \text{ and } p_1 = p_2 = q_1 = q_2 = 1/3.$$

The correction factor in this case is also $\lambda = 1/2$. Writing this one out, one has

$$\bar{X}_n = \bar{X}_{n-1} + \tfrac{1}{6}\{[F_0 + 2F_1 + 2F_2 + F_3]h$$
$$+ [G_0 + 2G_1 + 2G_2 + G_3]\,\Delta W_n\} \tag{3.5}$$

where

$$F_0 = f(t_{n-1}, \bar{X}_{n-1}),$$
$$F_1 = f\left(t_{n-1} + \tfrac{1}{2}h, \bar{X}_{n-1} + \tfrac{1}{2}F_0 h + \tfrac{1}{2}G_0 \Delta W_n\right),$$
$$F_2 = f\left(t_{n-1} + \tfrac{1}{2}h, \bar{X}_{n-1} + \tfrac{1}{2}F_1 h + \tfrac{1}{2}G_1 \Delta W_n\right),$$
$$F_3 = f\left(t_n, \bar{X}_{n-1} + F_2 h + G_2 \Delta W_n\right),$$

(3.6)

and the G_i's are defined as g evaluated at the same points. By expanding (3.6) and the corresponding expressions for the G_i's via Taylor one has

$$F_1 = f + \frac{1}{2}\left(\frac{\partial f}{\partial x}\right)gW(h) + O(h) + O(W^2(h)),$$

$$F_2 = f + \frac{1}{2}\left(\frac{\partial f}{\partial x}\right)gW(h) + O(h) + O(W^2(h)),$$

$$F_3 = f + \left(\frac{\partial f}{\partial x}\right)gW(h) + O(h) + O(W^2(h)),$$

$$G_1 = g + \frac{1}{2}\left[\frac{\partial g}{\partial t} + \left(\frac{\partial g}{\partial x}\right)f\right]h + \frac{1}{2}\left(\frac{\partial g}{\partial x}\right)gW(h)$$
$$+ \frac{1}{8}\left(\frac{\partial^2 g}{\partial x^2}\right)g^2 W^2(h)$$
$$+ O(h|W(h)|) + O(h^2) + O(|W(h)|^3),$$

$$G_2 = g + \frac{1}{2}\left[\frac{\partial g}{\partial t} + \left(\frac{\partial g}{\partial x}\right)f\right]h + \frac{1}{2}\left(\frac{\partial g}{\partial x}\right)gW(h)$$
$$+ \left[\frac{1}{4}\left(\frac{\partial g}{\partial x}\right)^2 g + \frac{1}{8}\left(\frac{\partial^2 g}{\partial x^2}\right)g^2\right]W^2(h)$$
$$+ O(h|W(h)|) + O(h^2) + O(|W(h)|^3),$$

(3.7)

and

$$G_3 = g + \left[\frac{\partial g}{\partial t} + \left(\frac{\partial g}{\partial x}\right)f\right]h + \left(\frac{\partial g}{\partial x}\right)gW(h)$$
$$+ \frac{1}{2}\left[\left(\frac{\partial g}{\partial x}\right)^2 g + \left(\frac{\partial^2 g}{\partial x^2}\right)g^2\right]W^2(h)$$
$$+ O(h|W(h)|) + O(h^2) + O(|W(h)|^3),$$

where the functions on the right above are evaluated at $(t_{n-1}, \bar{X}_{n-1})$ and, since the emphasis is on one step of the algorithm once again ΔW_n is replaced by $W(h)$. Substituting (3.7) into (3.5) and simplifying somewhat gives

$$
\begin{aligned}
\bar{X}_n = {} & \bar{X}_{n-1} + f(t_{n-1}, \bar{X}_{n-1})h \\
& + \frac{1}{2}\left[\left(\frac{\partial f}{\partial x}\right)g\right](t_{n-1}, \bar{X}_{n-1})hW(h) \\
& + g(t_{n-1}, \bar{X}_{n-1})W(h) \\
& + \frac{1}{2}\left[\frac{\partial g}{\partial t} + \left(\frac{\partial g}{\partial x}\right)f\right](t_{n-1}, \bar{X}_{n-1})hW(h) \\
& + \frac{1}{2}\left[\left(\frac{\partial g}{\partial x}\right)g\right](t_{n-1}, \bar{X}_{n-1})W^2(h) \\
& + \frac{1}{6}\left[\left(\frac{\partial g}{\partial x}\right)^2 g + \left(\frac{\partial^2 g}{\partial x^2}\right)g^2\right](t_{n-1}, \bar{X}_{n-1})W^3(h) \\
& + O(h^2) + O(hW^2(h)) + O(W^4(h)).
\end{aligned} \tag{3.8}
$$

In the quadratic mean squared sense the last three terms are all $O(h^4)$. Thus (3.8) indicates that the stochastic analogue of the classical fourth-order Runge-Kutta method is equivalent up to order h^4 in the mean square sense to the Mil'shtein scheme

$$
\bar{X}_n = \bar{X}_{n-1} + c_1 h + c_2 W(h) + c_3 W^2(h) + c_4 hW(h) + c_5 W^3(h) \tag{3.9}
$$

where the c_i have the functional forms

$$
c_1 = f, \qquad c_2 = g, \qquad c_3 = \frac{1}{2}\left(\frac{\partial g}{\partial x}\right)g,
$$

$$
c_4 = \frac{1}{2}\left[\left(\frac{\partial f}{\partial x}\right)g + \frac{\partial g}{\partial t} + \left(\frac{\partial g}{\partial x}\right)f\right], \tag{3.10}
$$

$$
c_5 = \frac{1}{6}\left[\left(\frac{\partial g}{\partial x}\right)^2 g + \left(\frac{\partial^2 g}{\partial x^2}\right)g^2\right],
$$

all evaluated at $(t_{n-1}, \bar{X}_{n-1})$. Rümelin's theorem indicates that the approximation (3.9), (3.10) converges to the solution of the Stratonovich equation

$$
dX = f(t, X)\, dt + g(t, X)\, dW \tag{3.11}
$$

or the Ito solution of

$$dX = \left[f + \frac{1}{2} \left(\frac{\partial g}{\partial x} \right) g \right] (t, X) \, dt + g(t, X) \, dW. \tag{3.12}$$

(The correction factor is $\lambda = 1/2$.) For the corresponding approximation of the form (3.9) to the Ito interpreted solution of (3.11), once again, one replaces f by $f - \frac{1}{2}(\partial g/\partial x)g$ in (3.10). In particular, one takes c_2, c_3, and c_5 as in (3.10),

$$c_1 = f - \frac{1}{2} \left(\frac{\partial g}{\partial x} \right) g,$$

$$\tag{3.13}$$

$$c_4 = \frac{1}{2} \left[\left(\frac{\partial f}{\partial x} \right) g + \frac{\partial g}{\partial t} + \left(\frac{\partial g}{\partial x} \right) f - \left(\frac{\partial g}{\partial x} \right)^2 g - \frac{1}{2} \left(\frac{\partial^2 g}{\partial x^2} \right) g^2 \right].$$

Summarizing, with these choices for the coefficients, the Mil'shtein scheme (3.9) is equivalent locally to the fourth-order Runge-Kutta method (3.5), (3.6) with f replaced by $f - \frac{1}{2} g (\partial g/\partial x)$ up to order $O(h^4)$ in the quadratic mean squared sense. The point of establishing this equivalence is to investigate the local truncation error arising from the implementation of this method along the lines of the discussion in the previous section. In particular, is there an improvement in the order of accuracy to $O(h^4)$?

Proceeding as in the last section, define $H = \Phi^2$ where, in this case, for an arbitrary $\bar{x}_{n-1}$, $\bar{w}_{n-1}$,

$$\Phi(t, x, w) = x - \bar{x}_{n-1} - c_1(t - t_{n-1})$$

$$- c_2(w - \bar{w}_{n-1}) - c_3(w - \bar{w}_{n-1})^2$$

$$- c_4(t - t_{n-1})(w - \bar{w}_{n-1}) - c_5(w - \bar{w}_{n-1})^3.$$

The question, then, is whether or not

$$E(H(t_n, X(t_n)) \mid X(t_{n-1}) = \bar{x}_{n-1}, W(t_{n-1}) = \bar{w}_{n-1})$$

$$= E \left(|X(t_n) - \bar{X}_n|^2 \mid X(t_{n-1}) = \bar{X}_{n-1} = \bar{x}_{n-1}, W(t_{n-1}) = \bar{w}_{n-1} \right)$$

$$= O(h^4). \tag{3.14}$$

If f and g have bounded derivatives through the sixth order, (3.14) results provided

$$LH(t_{n-1}, \bar{x}_{n-1}, \bar{w}_{n-1}) = 0,$$

$$L^2 H(t_{n-1}, \bar{x}_{n-1}, \bar{w}_{n-1}) = 0, \tag{3.15}$$

$$L^3 H(t_{n-1}, \bar{x}_{n-1}, \bar{w}_{n-1}) = 0,$$

following the argument given in the last section. Conditions (3.15) in turn require the vanishing of

$$L\Phi, \quad S\Phi, \quad LS\Phi, \quad SL\Phi, \quad S^2\Phi, \quad \text{and} \quad S^3\Phi$$

at $(t_{n-1}, \bar{x}_{n-1}, \bar{w}_{n-1})$, considering that

$$L^3 H = 2\left\{L(L\Phi)^2 + L(\Phi L^2\Phi) + L(S\Phi(SL\Phi + LS\Phi))\right\} + L(S^2\Phi)^2$$

$$= 2\left\{2L\Phi L^2\Phi + (SL\Phi)^2 + L\Phi L^2\Phi + \Phi L^3\Phi + S\Phi SL^2\Phi\right.$$

$$+ LS\Phi(SL\Phi + LS\Phi) + S\Phi(L(SL\Phi) + L^2(S\Phi))$$

$$+ \left. S^2\Phi(S^2(L\Phi) + S(LS\Phi))\right\} + 2S^2\Phi LS^2\Phi + (S^3\Phi)^2.$$

These requirements become

$$c_1 + c_3 = f(t_{n-1}, \bar{x}_{n-1}),$$

$$c_2 = g(t_{n-1}, \bar{x}_{n-1}),$$

$$c_4 + 3c_5 = \left[\frac{\partial g}{\partial t} + \left(\frac{\partial g}{\partial x}\right)f + \frac{1}{2}\left(\frac{\partial^2 g}{\partial x^2}\right)g^2\right](t_{n-1}, \bar{x}_{n-1}),$$

$$c_4 + 3c_5 = \left[\left(\frac{\partial f}{\partial x}\right)g\right](t_{n-1}, \bar{x}_{n-1}), \tag{3.16}$$

$$c_3 = \frac{1}{2}\left[\left(\frac{\partial g}{\partial x}\right)g\right](t_{n-1}, \bar{x}_{n-1}),$$

$$c_5 = \frac{1}{6}\left[\left(\frac{\partial g}{\partial x}\right)^2 g + \left(\frac{\partial^2 g}{\partial x^2}\right)g^2\right](t_{n-1}, \bar{x}_{n-1}).$$

For (3.16) to be consistent, it is necessary that

$$\frac{\partial g}{\partial t} + \left(\frac{\partial g}{\partial x}\right)f + \frac{1}{2}\left(\frac{\partial^2 g}{\partial x^2}\right)g^2 = \left(\frac{\partial f}{\partial x}\right)g \tag{3.17}$$

at the point $(t_{n-1}, \bar{x}_{n-1})$. In this case subtracting $3c_5$ from the right side of each of the middle equations in (3.16) and averaging leads to the form

$$c_4 = \frac{1}{2}\left[\left(\frac{\partial f}{\partial x}\right)g + \frac{\partial g}{\partial t} + \left(\frac{\partial g}{\partial x}\right)f - \left(\frac{\partial g}{\partial x}\right)^2 g - \frac{1}{2}\left(\frac{\partial^2 g}{\partial x^2}\right)g^2\right],$$

that is, (3.16) are identical to the choices (3.13). Note that, from (3.16), one has equivalently

$$
\begin{aligned}
c_4 &= \frac{\partial g}{\partial t} + f\left(\frac{\partial g}{\partial x}\right) - \frac{1}{2}g\left(\frac{\partial g}{\partial x}\right)^2 \\
&= L_X g - \frac{1}{2}\frac{\partial}{\partial x}\left(\frac{\partial g}{\partial x}g\right)g
\end{aligned}
\tag{3.18}
$$

where L_X is the Ito operator associated with the solution process X. The conclusion here is that only if (3.17) holds can truncation error $O(h^4)$ be achieved by a numerical scheme of the Mil'shtein type, and the situation is succintly summarized by the following result of Rümelin.

THEOREM 7.6   Suppose f and g have continuous and bounded partial derivatives up to the sixth order. Then:
(a)   The best numerical scheme for (3.14) has local truncation error $O(h^3)$ if and only if the coefficients of (2.1) do not satisfy (3.17). To obtain $O(h^3)$ it suffices to use the Heun scheme.
(b)   If (3.17) does hold, the best possible order is at least $O(h^4)$ which is realized by the fourth-order Runge-Kutta scheme with $\lambda = 1/2$.

EXAMPLE 7.2   Returning to Example 7.1

$$
dX = X\,dW
$$

note that (3.17) holds, since $f(t,x) = 0$, $g(t,x) = x$ makes both sides of (3.17) identically zero. One expects that the equivalent Runge-Kutta scheme (3.9), (3.13) will produce $O(h^4)$ one-step error order of magnitude in this case. Indeed, this scheme has the form here

$$
\begin{aligned}
\bar{X}_0 &= \bar{X} \\
\bar{X}_n &= \bar{X}_{n-1} - \tfrac{1}{2}\bar{X}_{n-1}h + \bar{X}_{n-1}\,\Delta W_n \\
&\quad + \tfrac{1}{2}\bar{X}_{n-1}(\Delta W_n)^2 - \tfrac{1}{2}\bar{X}_{n-1}h\,\Delta W_n + \tfrac{1}{6}\bar{X}_{n-1}(\Delta W_n)^3.
\end{aligned}
$$

In one step one has

$$
\bar{X}_1 = \bar{X}_0\left[1 - \tfrac{1}{2}h + W(h) + \tfrac{1}{2}W^2(h) - \tfrac{1}{2}hW(h) + \tfrac{1}{6}W^3(h)\right].
\tag{3.19}
$$

Expanding the corresponding exact solution value

$$
X(h) = X_0\exp\left\{-\tfrac{1}{2}h + W(h)\right\}
$$

yields

$$X(h) = X_0 \left[1 - \tfrac{1}{2}h + W(h) + \tfrac{1}{2} \left(-\tfrac{1}{2}h + W(h) \right)^2 + \tfrac{1}{6} \left(-\tfrac{1}{2}h + W(h) \right)^3 \right.$$
$$\left. + O((h + |W(h)|)^4) \right]$$
$$= X_0 \left[1 - \tfrac{1}{2}h + W(h) - \tfrac{1}{2}hW(h) + \tfrac{1}{2}W^2(h) + \tfrac{1}{6}W^3(h) \right.$$
$$\left. + O(h^2) + O(hW^2(h)) + O(W^4(h)) \right].$$

Comparing with (3.19), the one-step error order of magnitude, then, is

$$E\left(|X(h) - \bar{X}_1|^2 \mid X_0 = \bar{X}_0 = \bar{x}_0 \right)$$
$$= \bar{x}_0^2 E[O(h^2) + O(hW^2(h)) + O(W^4(h))]^2$$
$$= O(h^4).$$

A final remark about condition (3.17), which gives an indication how restrictive this condition is, seems appropriate. If $g \neq 0$, isolating $\partial g / \partial t$ and dividing by g^2 in (3.17) gives

$$\left(\frac{1}{g^2} \right) \frac{\partial g}{\partial t} = \frac{1}{g^2} \left[\left(\frac{\partial f}{\partial x} \right) g - \left(\frac{\partial g}{\partial x} \right) f \right] - \frac{1}{2} \left(\frac{\partial^2 g}{\partial x^2} \right)$$
$$= \frac{\partial}{\partial x} \left[\frac{f}{g} - \frac{1}{2} \left(\frac{\partial g}{\partial x} \right) \right]. \tag{3.20}$$

Hence the condition for reducibility (4.1.9) as given in Chapter 4, is satisfied. Recall that this means that there is a transformation

$$Y(t) = h(t, X(t)) \tag{3.21}$$

of the solution $X(t)$ which has the explicit representation

$$Y(t) = h(0, X(0)) + \int_0^t \bar{f}(s)\, ds + \int_0^t \bar{g}(s)\, dW(s) \tag{3.22}$$

for some functions $\bar{f}$ and $\bar{g}$ which can be determined. If, furthermore, h is invertible, that is, if there exists a function $k(t, x)$ such that

$$h(t, k(t, x)) = x \quad \text{and} \quad k(t, h(t, x)) = x,$$

then the formula

$$X(t) = k(t, Y(t)) \tag{3.23}$$

obtains an explicit representation of the solution of the original stochastic equation

$$dX = f(t, X)\, dt + g(t, X)\, dW. \tag{3.24}$$

In fact, in case (3.20) holds, one may take

$$h(t, x) = \int^x \frac{1}{g(t, u)}\, du,$$

$$\bar{f}(t) \equiv 0, \tag{3.25}$$

$$\bar{g}(t) \equiv 1$$

so that, if $\int^x (1/g)$ is invertible with inverse k, the solution of (3.24) can be written

$$X(t) = k(t, \int^{X_0} \frac{1}{g(0, u)}\, du + W(t)).$$

7.4   SYSTEMS OF STOCHASTIC EQUATIONS

In this section the discussion of numerical techniques is extended to the vector stochastic differential equation, or system of stochastic differential equations,

$$dX = f(t, X)\, dt + G(t, X)\, dW; \tag{4.1}$$

$f = \{f_i\}$ is an n-vector-valued function,

$G = \{g_{ij}\}$ is an $n \times m$-matrix-valued function,

$W = \{W_j\}$ is an m-dimensional Wiener process,

and the solution X is an n-dimensional diffusion process whose transition probabilities are determined by the drift vector f and the diffusion matrix GG^T. In the special case $m = 1$, the case of an n-dimensional system affected by a single scalar noise process, the techniques discussed in the first three sections can be applied to (the component equations of) (4.1) with similar results. So the case $m > 1$ is to be considered here particularly. It will prove convenient to write (4.1) in the form

$$dX = f(t, X)\, dt + \sum_{j=1}^{m} g_j(t, X)\, dW_j, \tag{4.1a}$$

where the g_j are the columns of the matrix G and the W_j are the independent scalar Wiener processes forming the components of W.

The scalar equivalent Heun scheme of the form

$$\bar{X}_n = \bar{X}_{n-1} + c_1(t_{n-1}, \bar{X}_{n-1})h + c_2(t_{n-1}, \bar{X}_{n-1})\,\Delta W_n$$

$$+ c_3(t_{n-1}, \bar{X}_{n-1})(\Delta W_n)^2$$

discussed in Section 7.2 suggests that one should try the form

$$\bar{X}_p = \bar{X}_{p-1} + c_0(t_{p-1}, \bar{X}_{p-1})h + \sum_{j=1}^{m} c_j(t_{p-1}, \bar{X}_{p-1})(\Delta W_j)_p$$

$$+ \sum_{j,k=1}^{m} c_{jk}(t_{p-1}, \bar{X}_{p-1})(\Delta W_j)_p(\Delta W_k)_p$$

(4.2)

where the $c_j = \{c_j^i\}$ and $c_{jk} = \{c_{jk}^i\}$ are n-vector valued functions, and

$$(\Delta W_j)_p = W_j(t_p) - W_j(t_{p-1}).$$

to obtain $O(h^3)$ accuracy in this case. However, it follows that (4.2) leads to the desired accuracy only if the symmetry condition

$$(\nabla_x g_i)g_j = (\nabla_x g_j)g_i$$

(4.3)

holds for all $i, j = 1, \ldots, m$. The parameter choices

$$c_0 = \left[f - \frac{1}{2}\sum_{j=1}^{m}(\nabla_x g_j)g_j \right](t_{p-1}, \bar{X}_{p-1})$$

(4.4a)

$$c_j = g_j(t_{p-1}, \bar{X}_{p-1})$$

(4.4b)

$$c_{jk} + c_{kj} = [(\nabla_x g_j)g_k](t_{p-1}, \bar{X}_{p-1})$$

(4.4c)

must be made to effect $O(h^3)$ one-step accuracy when (4.3) holds. Indeed the requirement (4.3) can be seen by observing that the multidimensional analogue to the last equation in (2.29) is

$$S_k(S_{ij}\Phi_i)(t_{p-1}, \bar{x}_{p-1,i}, \bar{w}_{p-1,1}, \ldots, \bar{w}_{p-1,m}) = 0$$

(4.5)

for all $i = 1, \ldots, n$, $j, k = 1, \ldots, m$, where

$$\bar{x}_{p-1} = (\bar{x}_{p-1,1}, \ldots, \bar{x}_{p-1,n}) \quad \text{and} \quad \bar{w}_{p-1} = (\bar{w}_{p-1,1}, \ldots, \bar{w}_{p-1,m})$$

are arbitrary and

$$\Phi_i(t, x_i, w_1, \ldots, w_m) = x_i - \bar{x}_{p-1,i} - c_0^i(t - t_{p-1})$$

$$-\sum_{j=1}^{m} c_j^i(w_j - \bar{w}_{p-1,j}) - \sum_{j,k=1}^{m} c_{jk}^i(w_j - \bar{w}_{p-1,j})(w_k - \bar{w}_{p-1,k}),$$

$$S_{ij}\Phi = g_{ij}\frac{\partial \Phi}{\partial x_i} + \frac{\partial \Phi}{\partial w_j}, \tag{4.6}$$

and

$$S_k\Phi = \sum_{i=1}^{n} g_{ik}\frac{\partial \Phi}{\partial x_i} + \frac{\partial \Phi}{\partial w_k}.$$

Specifically, (4.5) has the form

$$\sum_{\nu=1}^{n} g_{\nu k}\frac{\partial g_{ij}}{\partial x_\nu} - (c_{jk}^i + c_{kj}^i) = 0, \tag{4.7}$$

and so the symmetry of the second term in (4.7) forces symmetry of the first in j and k, that is,

$$\sum g_{\nu k}\frac{\partial g_{ij}}{\partial x_\nu} = \sum g_{\nu j}\frac{\partial g_{ik}}{\partial x_\nu}. \tag{4.8}$$

In vector-matrix form (4.8) becomes (4.3).

Summarizing this section one has the Rümelin (1982) theorem:

THEOREM 7.7   Suppose f_i and g_{ij} have continuous and bounded derivatives up to the fourth order. Then:

(a)   The best numerical scheme for (4.1) has local truncation error $O(h^2)$ if and only if the matrix G does not satisfy (4.3). To obtain $O(h^2)$ it suffices to use the Euler scheme (4.2) with c_0 and c_j, $1 \leq j \leq m$, given by (4.4a and b) and $c_{jk} = 0$.

(b)   If (4.3) does hold, the best possible order is at least $O(h^3)$, which is realized by the equivalent Heun scheme (4.2), (4.4).

EXAMPLE 7.2   If $m = n$ and

$$G(x) = \text{diag}\{g_{11}(x_1), g_{22}(x_2), \ldots, g_{nn}(x_n)\},$$

then G satisfies (4.3). Hence, when each component equation is subject to an independent noise term with decoupled intensity, the system may be handled similarly to the scalar case. The stochastic food chain, Example 6.3, has this form.

7.5 REMARKS ON METHODS GENERATED BY ITERATED STOCHASTIC INTEGRALS

Numerical methods for stochastic differential equations are known which are free of the truncation error limitation discussed in Sections 7.3 and 7.4. A number of these methods exploit the possibility of numerically simulating mixed multiple or iterated Wiener integrals such as

$$\int_0^h \int_0^t \int_0^s r \, dW(r) \, dW(s) \, dt, \tag{5.1}$$

in addition to the Wiener increment

$$W(h) = \int_0^h dW(t). \tag{5.2}$$

For example Mil'shtein (1974) suggested dealing with the problem discussed in Section 7.3 by adding a term of the form

$$\int_0^h W(t) \, dt = \int_0^h \int_0^t dW(s) \, dt \tag{5.3}$$

to the scheme (3.9). Such schemes can be derived via a stochastic Taylor formula for the true solution in which iterated stochastic integrals arise. [See, for example, Wagner and Platen (1978), and Platen (1980, 1981); in particular, the notation adopted below and the basic proposition given appear in Wagner and Platen (1978).] As an illustration, in this section the above-mentioned Mil'shtein algorithm is derived. This scheme possesses the same error order of magnitude as the stochastic Runge-Kutta scheme given in Section 7.3 without being subject to the restriction (3.17) relating drift and diffusion coefficients. Here, however, implementation necessitates that a second random variable, namely (5.3), must be simulated: the scheme requires, at each step, producing sample values of the random variables $Z_1 \sim W(h)$ and $Z_2 \sim \int_0^h W(t) \, dt$ where Z_1 is $N(0, h)$, Z_2 is $N(0, h^3/3)$, and $EZ_1 Z_2 = h^2/2$.

To begin, some notation is introduced which permits convenient writing of mixed multiple stochastic integrals involving a single scalar Wiener process W. For k a positive integer, let $\alpha = (j_1, \ldots, j_k)$ represent a k-tuple of zeroes and ones; define $l(\alpha) = k$ and $n(\alpha) =$ number of zeroes appearing in α. Also define the differential

$$dY^{(j)}(u) = \begin{cases} du & \text{if } j = 0, \\ dW(u) & \text{if } j = 1. \end{cases}$$

For a random function F defined on an interval $[t_0, T]$ and any such $\alpha = (j_1, \ldots, j_k)$ denote the corresponding mixed multiple or iterated stochastic integral by

$$I_\alpha(F; t_0, T) = \int_{t_0}^{T} \int_{t_0}^{u_k} \cdots \int_{t_0}^{u_2} F(u_1)\, dY^{(j_1)}(u_1) \cdots$$

$$dY^{(j_{k-1})}(u_{k-1})\, dY^{(j_k)}(u_k) \quad (5.4)$$

whenever it exists; for $F \equiv 1$, write $I_\alpha(t_0, T)$. Throughout integrands in (5.4) will have the specific form

$$F(\cdot) = F(\cdot, X(\cdot))$$

where X denotes a solution of a stochastic differential equation. The following proposition gives the basic properties of such integrals and can be viewed as an extension of Theorem 2.6 in Chapter 2. Its proof is left as an exercise.

THEOREM 7.8   Suppose the function F is in the class $\mathcal{L}^2$ (see Section 2.2) on $[t_0, T]$. Then

(a) $E(I_\alpha(F; t_0, T)) = 0$ if $j_i = 1$, at least some i, $1 \le i \le k$.

(b) $E(I_\alpha^2(F; t_0, T)) = O((T - t_0)^{l(\alpha) + n(\alpha)})$.

Note that the condition given in part (a) of the theorem simply assures that I_α is a stochastic integral; (a) implies that $I_\alpha(F; t_0, t)$, $t \in [t_0, T]$, is a martingale with respect to an appropriate family of σ-algebras of events in the underlying probability space (similarly to Theorem 2.7, Section 2.2).

As examples of the notation and application of Theorem 7.8 consider the stochastic integrals (5.1), (5.2), and (5.3): they are denoted by $I_{(0,1,1,0)}(0, h)$, $I_{(1)}(0, h)$, and $I_{(1,0)}(0, h)$, respectively, they all have expected value zero, and they have quadratic mean squared orders $O(h^6)$, $O(h)$, and $O(h^3)$, respectively.

Now, the first two cases of the stochastic Taylor formula, alluded to above, are given explicitly for the expansion of the solution $X(t)$ of the stochastic initial value problem

$$dX = f(t, X)\, dt + g(t, X)\, dW \qquad\qquad (5.5)$$

$$X(t_0) = X_0 \qquad\qquad (5.6)$$

on the interval $[t_0, T]$. The idea here is that, upon writing the integral equation equivalent

$$X(t) = X_0 + \int_{t_0}^{t} f(s, X(s))\, ds + \int_{t_0}^{t} g(s, X(s))\, dW(s), \qquad (5.7)$$

one may expand $f(s, X(s))$ and $g(s, X(s))$ by Ito's formula if f and g are sufficiently smooth. It follows that if f and g have continuous second partial derivatives,

$$
\begin{aligned}
X(t) = X_0 &+ \int_{t_0}^t \left\{ f(t_0, X_0) + \int_{t_0}^s Lf(r, X(r))\, dr \right. \\
&+ \left. \int_{t_0}^s \left[\frac{\partial f}{\partial x} g \right](r, X(r))\, dW(r) \right\} ds \\
&+ \int_{t_0}^t \left\{ g(t_0, X_0) + \int_{t_0}^s Lg(r, X(r))\, dr \right. \\
&+ \left. \int_{t_0}^s \left[\frac{\partial g}{\partial x} g \right](r, X(r))\, dW(r) \right\} dW(s)
\end{aligned}
\tag{5.8}
$$

(where L is the Ito operator associated with (5.5), as usual). Making use of the notation introduced above one can write (5.7) and (5.8), respectively, as

$$
X(t) = X_0 + I_{(0)}(f; t_0, t) + I_{(1)}(g; t_0, t)
$$

and

$$
\begin{aligned}
X(t) = X_0 &+ f(t_0, X_0)[t - t_0] + g(t_0, X_0)[W(t) - W(t_0)] \\
&+ I_{(0,0)}(Lf; t_0, t) + I_{(1,0)}\left(\frac{\partial f}{\partial x} g; t_0, t \right) \\
&+ I_{(0,1)}(Lg; t_0, t) + I_{(1,1)}\left(\frac{\partial g}{\partial x} g; t_0, t \right).
\end{aligned}
$$

The sum of the four integrals in (5.8) can be considered as constituting the remainder $R_1(t, t_0)$ of the stochastic Taylor formula

$$
X(t) = X_0 + f(t_0, X_0)[t - t_0] + g(t_0, X_0)[W(t) - W(t_0)] + R_1(t, t_0).
\tag{5.9}
$$

If the integrands of the double integrals appearing in R_1 are bounded (mean square bounded is sufficient) on $[t_0, T]$, then $R_1(t, t_0) = O((t - t_0)^2)$ in the quadratic mean squared sense. This follows since $I_{(0,0)}(Lf; t_0, t)$, being an ordinary double integral, has quadratic mean squared order $O((t - t_0)^4)$, and the remaining integrals have orders $O((t - t_0)^3)$, $O((t - t_0)^3)$, and $O((t - t_0)^2)$, respectively, according to Theorem 7.8. Consequently, the

Euler scheme, given in Section 7.1,

$$\bar{X}_0 = X_0$$
$$\bar{X}_n = \bar{X}_{n-1} + f(t_{n-1}, \bar{X}_{n-1})\Delta t_n + g(t_{n-1}, \bar{X}_{n-1})\Delta W_n \qquad (5.10)$$
$$n = 1,\ldots,N$$

corresponding to the partition $t_0 < t_1 < \cdots < t_N = T$, where $\Delta t_n = t_n - t_{n-1}$ and $\Delta W_n = W(t_n) - W(t_{n-1})$, for obtaining an approximate solution is verified to yield local error $O(h^2)$, $h = \max_n \Delta t_n$; the result is the same as that obtained in Section 7.1.

Continuing, one may expand each of the integrands in the double integrals in (5.8) via Ito's formula, provided f and g are sufficiently smooth, to obtain (denoting evaluation at (t_0, X_0) by the subscript 0):

$$X(t) = X_0 + f_0(t - t_0) + g_0(W(t) - W(t_0))$$
$$+ \left[\frac{\partial g}{\partial x}g\right]_0 I_{(1,1)}(t,t_0) + R_2(t,t_0) \qquad (5.11)$$

where

$$R_2(t,t_0) = \tfrac{1}{2}[Lf]_0(t-t_0)^2 + \left[\frac{\partial f}{\partial x}g\right]_0 I_{(1,0)}(t,t_0)$$

$$+ [Lg]_0 I_{(0,1)}(t,t_0)$$

$$+ I_{(0,0,0)}(L^2 f; t_0, t) + I_{(1,0,0)}\left(\frac{\partial}{\partial x}(Lf)g; t_0, t\right)$$

$$+ I_{(0,1,0)}\left(L\left(\frac{\partial f}{\partial x}g\right); t_0, t\right)$$

$$+ I_{(1,1,0)}\left(\frac{\partial}{\partial x}\left(\frac{\partial f}{\partial x}g\right)g; t_0, t\right)$$

$$+ I_{(0,0,1)}(L^2 g; t_0, t) + I_{(1,0,1)}\left(\frac{\partial}{\partial x}(Lg)g; t_0, t\right)$$

$$+ I_{(0,1,1)}\left(L\left(\frac{\partial g}{\partial x}g\right); t_0, t\right)$$

$$+ I_{(1,1,1)}\left(\frac{\partial}{\partial x}\left(\frac{\partial g}{\partial x}g\right)g; t_0, t\right). \qquad (5.12)$$

Assuming, once again, boundedness of the integrands, Theorem 7.8 can be applied to verify that

$$R_2(t, t_0) = O((t - t_0)^3)$$

in the quadratic mean squared sense. Equation (5.11) constitutes the second case of the stochastic Taylor formula. Note that the integral $I_{(1,1)}$ appearing in (5.11) may be calculated:

$$I_{(1,1)}(t, t_0) = \int_{t_0}^t [W(s) - W(t_0)] \, dW(s)$$

$$= \tfrac{1}{2} \left\{ [W(t) - W(t_0)]^2 - (t - t_0) \right\},$$

so that (5.11) may be written as:

$$X(t) = X_0 + \left(f_0 - \frac{1}{2} \left[\frac{\partial g}{\partial x} g \right]_0 \right) [t - t_0] + g_0[W(t) - W(t_0)]$$

$$+ \frac{1}{2} \left[\frac{\partial g}{\partial x} g \right]_0 [W(t) - W(t_0)]^2 + R_2(t, t_0). \qquad (5.13)$$

Equation (5.13) indicates that Heun scheme, therefore,

$$\bar{X}_n = \bar{X}_{n-1} + \left[f - \frac{1}{2} \frac{\partial g}{\partial x} g \right]_{n-1} \Delta t_n$$

$$+ g_{n-1} \Delta W_n + \frac{1}{2} \left[\frac{\partial g}{\partial x} g \right]_{n-1} (\Delta W_n)^2$$

(where $n - 1$ subscripts indicate evaluation at $(t_{n-1}, \bar{X}_{n-1})$) should experience local error order of magnitude $O(h^3)$; again this is the same result as obtained in Section 7.2.

The situation changes at the next level of iteration of this scheme. Similarly, the triple integrals in (5.12) are expanded via Ito's formula and terms of quadratic mean squared order $O((t - t_0)^3)$ in (5.11) are separated out; by inspection and another application of Theorem 7.8 it becomes clear that the $O((t - t_0)^3)$ terms in (5.12) are precisely the $I_{(1,0)}$ and $I_{(0,1)}$ terms together with the first term in the expansion of the $I_{(1,1,1)}$ triple integral. One obtains the formula

$$X(t) = X_0 + f_0(t - t_0) + g_0(W(t) - W(t_0))$$

$$+ \left[\frac{\partial g}{\partial x} g \right]_0 I_{(1,1)}(t, t_0)$$

$$+ \left[\frac{\partial f}{\partial x} g\right]_0 I_{(1,0)}(t, t_0) + [Lg]_0 I_{(0,1)}(t, t_0)$$

$$+ \left[\frac{\partial}{\partial x}\left(\frac{\partial g}{\partial x} g\right) g\right]_0 I_{(1,1,1)}(t, t_0) + R_3(t, t_0). \qquad (5.14)$$

Here R_3 consists of double, triple, and quadruple mixed-type stochastic integral terms and under the appropriate boundedness assumptions

$$R_3(t, t_0) = O((t - t_0)^4)$$

obtains, once again, from Theorem 7.8. Now the following relations can be used to simplify the stochastic integrals appearing in (5.14):

$$I_{(1,1)}(t, t_0) = \tfrac{1}{2}\left\{[W(t) - W(t_0)]^2 - (t - t_0)\right\}$$

$$I_{(0,1)}(t, t_0) = (t - t_0)(W(t) - W(t_0)) - I_{(1,0)}(t, t_0)$$

$$\int_{t_0}^t [W(s) - W(t_0)]^2 \, dW(s) = \tfrac{1}{3}[W(t) - W(t_0)]^3 - I_{(1,0)}(t, t_0).$$

Using these relations specifically to evaluate the first, third, and fourth integrals in (5.14) results in

$$X(t) = X_0 + \left[f - \frac{1}{2}\frac{\partial g}{\partial x} g\right]_0 (t - t_0) + g_0[W(t) - W(t_0)]$$

$$+ \left[Lg - \frac{1}{2}\frac{\partial}{\partial x}\left(\frac{\partial g}{\partial x} g\right) g\right]_0 (t - t_0)[W(t) - W(t_0)]$$

$$+ \frac{1}{2}\left[\frac{\partial g}{\partial x} g\right]_0 [W(t) - W(t_0)]^2$$

$$+ \frac{1}{6}\left[\frac{\partial}{\partial x}\left(\frac{\partial g}{\partial x} g\right) g\right]_0 [W(t) - W(t_0)]^3$$

$$+ \left[\frac{\partial f}{\partial x} g - Lg\right]_0 I_{(1,0)}(t, t_0) + R_3(t, t_0). \qquad (5.15)$$

Equation (5.15) indicates that the numerical scheme

$$\bar{X}_n = \bar{X}_{n-1} + \left[f - \frac{1}{2}\frac{\partial g}{\partial x} g\right]_{n-1} \Delta t_n + g_{n-1}\Delta W_n$$

$$+ \left[Lg - \frac{1}{2}\frac{\partial}{\partial x}\left(\frac{\partial g}{\partial x} g\right) g\right]_{n-1} \Delta t_n \Delta W_n + \frac{1}{2}\left[\frac{\partial g}{\partial x} g\right]_{n-1} (\Delta W_n)^2$$

$$+ \frac{1}{6}\left[\frac{\partial}{\partial x}\left(\frac{\partial g}{\partial x} g\right) g\right]_{n-1} (\Delta W_n)^3$$

$$+ \left[\frac{\partial f}{\partial x}g - Lg\right]_{n-1} I_{(1,0)}(t_{n-1}, t_n) \tag{5.16}$$

has local error $O(h^4)$. Notice that if

$$\frac{\partial f}{\partial g}g = Lg \tag{5.17}$$

the algorithm (5.16) reduces to the method given in Section 7.3 analogous to the classical fourth-order Runge-Kutta scheme. Equation (5.17) is precisely the consistency condition (3.17) required to validate the error order of magnitude claim in this case.

Higher order of magnitude schemes can be obtained at the expense of needing to simulate more iterated stochastic integrals upon implementation. Finally, derivative-free schemes can be obtained for all methods discussed in this chapter by replacing the derivatives that appear by differences of the appropriate order. Also the recent availability of "supercomputer" vector and parallel processors makes the implementation of these numerical methods more feasible.

EXERCISES

7.1 Verify Eqs. (2.28a) and (2.28b).

7.2 Redo the analysis in the example at the end of Section 7.2 for the equation

$$dX = \alpha X \, dt + \beta X \, dW$$

where α and β are constants.

7.3 Prove that the Heun method has global truncation error order of magnitude $O(h^2)$ in the quadratic mean squared sense.

7.4 For what choices of α and β will the fourth-order Runge-Kutta scheme (3.9), (3.13) for the equation

$$dX = \alpha X \, dt + \beta B \, dW$$

produce one-step error order of magnitude $O(h^4)$?

7.5 (a) Show that the stochastic equation

$$dX = (1 + X^2)(X \, dt + dW)$$

satisfies (3.17). Use the method of Chapter 4 to solve this equation.

   (b) Write out the equivalent fourth order Runge-Kutta scheme (3.9), (3.13) for the equation in part (a).

(c)   Following the example at the end of Section 7.2, expand the solution in part (a) by Taylor and compare $X(h)$ with $\bar{X}_1$, the Runge-Kutta approximate from part (b), to verify $O(h^4)$ order of magnitude for the one-step error.

7.6 Verify (4.5) and (4.7).

7.7 Show that (4.3) holds for any system forced by a single scalar Wiener process. (What is G in this case?)

7.8 State conditions on f and g under which the schemes (2.14) and (3.9), (3.13) are numerically stable. Verify your answer.

7.9 Verify that

$$\int_0^h W(t)\, dt$$

is $\mathcal{N}(0, h^3/3)$ and that

$$E\left(W(h) \int_0^h W(t)\, dt \right) = h^2/2.$$

7.10 Prove Theorem 7.8.

7.11 Write out an explicit expression for the remainder $R_3(t, t_0)$ in (5.14).

References

Arnold, L. (1974). *Stochastic Differential Equations: Theory and Applications*, Wiley, New York.

Ash, R. B. (1972). *Analysis and Probability*, Academic Press, New York.

Barra, M., G. Del Grosso, A. Gerardi, G. Koch, and F. Marchetti (1979). Some Basic Properties of Stochastic Population Models, *Systems Theory in Immunology*, Lect. Notes in Biomath. Vol. 32, Springer-Verlag, Berlin, pp. 165–174.

Bartle, R. G. (1966). *The Elements of Integration*, Wiley. New York.

Bharucha-Reid, A. T. (1960). *Elements of the Theory of Markov Processes*, McGraw-Hill, New York.

Brauman, C. A. (1983). Population growth in random environments, *Bull. Math. Biol. 45*, 635–641.

Bucy, R. S. (1965). Stability and positive supermartingales, *J. Diff. Eq. 1*, 151–155.

Chandrasekhar, S. (1943). Stochastic problems in physics and astronomy, *Rev. Mod. Phys. 15*, 1–89.

Chesson, P. (1978). Prey-predator theory and variability, *Ann. Rev. Ecol. Syst. 9*, 323–347.

Chesson, P. (1982). The stabilizing effect of a random environment, *J. Math. Biol. 15*, 1–36.

Conway, E. (1971). Stochastic equations with discontinuous drift, *Trans. Amer. Math. Soc. 157*, 235–245.

Doob, J. L. (1953). *Stochastic Processes*, Wiley, New York.

Doob, J. L. (1955). Martingales and one-dimensional diffusion, *Trans. Amer. Math. Soc.* **78**, 168–208.

Doos, H. (1977). Liens entre equations differentials stochastiques et ordinaires, *Ann. Inst. H. Poincare* **13**, 99–125.

Dynkin, E. B. (1965). *Markov Processes*, Vol. 1, 2, Springer-Verlag, Berlin.

Feller, W. (1954). Diffusion processes in one dimension, *Trans. Amer. Math. Soc.* **77**, 1–31.

Feller, W. (1968/1971). *An Introduction to Probability Theory and its Applications*, Vols. 1, 2, 3rd ed., Wiley, New York.

Freedman, H. I. *Deterministic Mathematical Models in Population Ecology*, Marcel Dekker, New York.

Freidlin, M. I. (1969). Markov processes and differential equations, Progress in Mathematics Vol. 3, Plenum, New York, pp. 1–55.

Friedman, A. (1972). Stability and angular behavior of solutions of stochastic differential equations, Lect. Notes in Math. Vol. 294, Springer-Verlag, Berlin, pp. 14–20.

Friedman, A., and M. Pinsky (1973). Asymptotic behavior of solutions of linear stochastic differential systems, *Trans. Amer. Math. Soc.* **181**, 1–22.

Friedman, A., and M. Pinsky (1973). Asymptotic behavior and spiraling properties of stochastic equations, *Trans. Amer. Math. Soc.* **186**, 331–358.

Friedman, A. (1975). *Stochastic Differential Equations and Applications*, Vols. 1, 2, Academic Press, New York.

Gard, T. C. (1976). A general uniqueness theorem for stochastic differential equations, *SIAM J. Cont. Opt.* **14**, 445–457.

Gard, T. C. (1977). Rate of decay for solutions of stochastic differential equations, *J. Math. Anal. Appl.* **61**, 142–150.

Gard, T. C. (1978). Pathwise uniqueness for solutions of systems of stochastic differential equations, *Stoc. Proc. Appl.* **6**, 253–260.

Gard, T. C. (1984). Persistence in stochastic food web models, *Bull. Math. Biol.* **46**, 357–370.

Gard, T. C. (1986). Stability for multispecies population models in random environments, *J. Nonlinear Anal.: Theory, Meth.,Appl.* **10**, 1411–1419.

Gard, T. C., and D. Kannan (1976). On a stochastic differential equation modeling of prey-predator evolution, *J. Appl. Prob.* **13**, 429–443.

Gelfand, I. M. (1964). *Generalized Functions*, Academic Press.

Gihman, I. I., and A. V. Skorohod (1969). *Introduction to the Theory of Random Processes*, Saunders, Philadelphia.

Gihman, I. I., and A. V. Skorohod (1972). *Stochastic Differential Equations*, Springer-Verlag, New York.

Gihman, I. I., and A. V. Skorohod (1974/1975/1979). *The Theory of Stochastic Process*, Vol. 1, 2, 3, Springer-Verlag, New York.

Girsanov, I. V. (1962). An example of nonuniqueness of the solution of K. Ito's stochastic equation, *Theor. Prob. Appl. 1*, 325–331.

Goel, N. S., S. C. Maitra, and E. W. Montroll (1971). On the Volterra and other nonlinear models of interacting populations, *Rev. Mod. Phys. 43*, 231–276.

Goel, N. S., and N. Richter-Dyn (1974). *Stochastic Models in Biology*, Academic Press, New York.

Goh, B. S. (1977). Global stability in many species systems, *Amer. Natur. 111*, 135–143.

Goldstein, J. A. (1969). Second order Ito processes, *Nagoya Math. J. 36*, 27–63.

Gray, A. H., and T. K. Caughey (1965). A controversy in problems involving random parametric excitation, *J. Math. Phys. 44*, 288–296.

Greenside, H. S., and E. Helfand (1981). Numerical integration of stochastic differential equations—II, *Bell Sys. Tech. J. 60*, 1927–1940.

Hartman, P. (1973). *Ordinary Differential Equations*, Hartman, Baltimore (orig. ed. Wiley, New York, 1964).

Has'minskii, R. Z. (1962). On the stability of the trajectory of Markov processes, *J. Appl. Math. Mech. 26*, 1554–1565.

Has'minskii, R. Z. (1965). On equations with random perturbations, *Theor. Prob. Appl. 19*, 361–364.

Has'minskii, R. Z. (1966). On the stability of nonlinear stochastic systems, *J. Appl. Math. Mech. 30*, 1082–1089.

Has'minskii, R. Z. (1967). Necessary and sufficient conditions for the asymptotic stability of linear stochastic systems, *Theor. Prob. Appl. 12*, 144–147.

Has'minskii, R. Z. (1967). Stability in the first approximation for stochastic systems, *J. Appl. Math. Mech. 31*, 1025–1030.

Has'minskii, R. Z. (1980). *Stochastic Stability of Differential Equations*, Sijthoff and Noordhoff, Alphen aan den Rijn, Netherlands.

Helfand, E. (1979). Numerical integration of stochastic differential equations, *Bell Sys. Tech. J. 58*, 2289–2299.

Ikeda, N., and S. Watanabe (1981). *Stochastic Differential Equations and Diffusion Processes*, North Holland, New York.

Ito, K. (1951). *On Stochastic Differential Equations*, Amer. Math. Soc. Memoirs No. 4, New York.

Ito, K. (1961). *Lectures on Stochastic Processes*, Tata Inst. Fund. Research, Bombay.

Ito, K., and H. P. McKean, (1965). *Diffusion Processes and Their Sample Paths*, Springer-Verlag, Berlin.

Kannan, D. (1979). *An Introduction to Stochastic Processes*, North Holland, New York.

Kesten, H., and Y. Ogura (1981). Recurrence properties of Lotka-Volterra models with random fluctuations, *J. Math. Soc. Japan 32*, 335–366.

Klauder, J. R., and W. P. Petersen (1985). Numerical integration of multiplicative-noise stochastic differential equations, *SIAM J. Num. Anal 22*, 1153–1166.

Kliemann, W. (1983). Qualitative theory of stochastic dynamical systems—applications to life sciences, *Bull. Math. Biol. 45*, 483–506.

Kloeden, P. E., and R. A. Pearson (1977). The numerical solution of stochastic differential equations, *J. Austral. Math. Soc. Ser. B, 20*, 8–12.

Kozin, F. (1969). A survey of stability of stochastic systems, *Automatica 5*, 95–112.

Krener, A. J. (1979). A formal approach to stochastic integration and differential equations, *Stochastics 3*, 105–125.

Krener, A. J., and C. Lobry (1981). The complexity of stochastic differential equations, *Stochastics 4*, 193–203.

Kunita, H. (1980). On the representation of solutions of stochastic differential equations, *Seminaire Probabilities XIV*, Lect. Notes in Math. Vol. 784, Springer-Verlag, Berlin, pp. 282–304.

Kushner, H. J. (1967). Converse theorems for stochastic Lyapunov functions, *SIAM J. Cont. 5*, 228–233.

Kushner, H. J. (1967). *Stochastic Stability and Control*, Academic Press, New York.

Kushner, H. J. (1972). Stability and the existence of diffusions with discontinuous or rapidly growing drift terms, *J. Diff. Eq. 11*, 156–168.

Kushner, H. J. (1974). On the weak convergence of interpolated Markov chains to a diffusion, *Ann. Prob. 2*, 40–50.

Kushner, H. J. (1977). *Probability Methods for Approximations in Stochastic Control and for Elliptic Equations*, Academic Press, New York.

Kussmaul, A. U. (1977). *Stochastic Integration and Generalized Martingales*, Pitman, London.

Lamperti, J. (1964). A simple construction of certain diffusion processes, *J. Math. Kyoto Univ. 4*, 647–668.

Leibowitz, M. A. (1963). Statistical behavior of linear systems with randomly varying parameters, *J. Math. Phys. 4*, 852–858.

Loeve, M. (1977/1978). *Probability Theory*, Vols. I, II, Springer-Verlag, New York.

Ludwig, D. (1974). *Stochastic Population Theories*, Lect. Notes in Biomath. Vol. 3, Springer-Verlag, Berlin.

Maistrov, L. E. (1974). *Probability Theory: A Historical Sketch*, Academic Press, New York.

Maruyama, G. (1955). Continuous Markov processes and stochastic equations, *Rend. Circ. Mat. Palermo 4*, 48–90.

McKean, H. P. (1969). *Stochastic Integrals*, Academic Press, New York.

McShane, E. J. (1974). *Stochastic Calculus and Stochastic Models*, Academic Press, New York.

Mil'shtein, G. N. (1974). Approximate integration of stochastic differential equations, *Theor. Prob. Appl. 19*, 557–562.

Mil'shtein, G. N. (1978). A method of second-order accuracy integration of stochastic differential equations, *Theor. Prob. Appl. 23*, 396–401.

Mortensen, R. E. (1969). Mathematical problems of modeling stochastic nonlinear dynamical systems, *J. Statist. Phys. 1*, 271–296.

Narita, K. (1982). No explosion criteria for stochastic differential equations, *J. Math. Soc. Japan 34*, 191–203.

Narita, K. (1982). Remarks on nonexplosion theorem for stochastic differential equations, *Kodai Math. J. 5*, 395–401.

Narita, K. (1983). A priori estimate and asymptotic behavior for solutions of stochastic differential equations, *Yokohama J. Math. 30*, 91–101.

Nisbet, R. M., and W. S. C. Gurney (1982). *Modelling Fluctuating Populations*, Wiley, New York.

Okabe, Y., and A. Shimizu (1975). On the pathwise uniqueness of solutions of stochastic differential equations, *J. Math. Kyoto Univ. 15*, 455–466.

Okamura, H. (1942). Condition necessaire et suffisante remplie par les equations differentielles ordinaires sans points de Peano, *Mem. Coll. Sci. Kyoto Univ. A24*, 21–28.

Papanicolaou, G. C., and W. Kohler (1974). Asymptotic theory of mixing stochastic ordinary differential equations. *Comm. Pure Appl. Math. 27*, 641–668.

Pinsky, M. A. (1974). Stochastic stability and the Dirichlet problem, *Comm. Pure Appl. Math. 27*, 311–350.

Platen, E. (1980). Weak convergence of approximations to Ito integral equations, *Z. Angew. Math. Mech. 60*, 609–614.

Platen, E. (1981). An approximation method for a class of Ito processes, *Lith. Math. J. 21*, 121–133.

Polansky, P. (1979). Invariant distributions for multipopulation models in random environments, *Theor. Pop. Biol. 16*, 25–34.

Prohorov, Yu. V., and Yu. A. Rozanov (1969). *Probability Theory*, Springer-Verlag, Berlin.

Protter, P. E. (1977). On the existence, uniqueness, convergence, and explosions of solutions of stochastic integral equations, *Ann. Prob. 5*, 243–261.

Rao, N. J., J. D. Borwankar, and D. Ramkrishma (1974). Numerical solution of Ito integral equations, *SIAM J. Cont. 12*, 124–139.

Ricciardi, L. M. (1974). *Diffusion Processes and Related Topics in Biology*, Lect. Notes in Biomath. Vol. 14, Springer-Verlag, Berlin.

Rouche, N., P. Habets, and M. Laloy (1977). *Stability Theory by Lyapunov's Direct Method*, Springer-Verlag, New York.

Royden, H. L. (1968). *Real Analysis*, 2nd ed., Macmillan, New York.

Rudin, W. (1976). *Principles of Mathematical Analysis*, 3rd ed. McGraw-Hill, New York.

Rümelin, W. (1982). Numerical treatment of stochastic differential equations, *SIAM J. Num. Anal. 19*, 604–613.

Saaty, T. L. (1967). *Modern Nonlinear Equations*, McGraw-Hill, New York.

Samuels, J. C. (1960). On the stability of random systems and the stabilization of deterministic systems, *J. Acoust. Soc. Amer. 32*, 594–601.

Schoener, T. W. (1973). Population growth regulated by intraspecific competition for energy or time: some simple representations, *Theor. Pop. Biol. 4*, 56–84.

Schuss, Z. (1980). *Theory and Applications of Stochastic Differential Equations*, Wiley, New York.

Skorohod, A. V. (1958). Limit theorems for Markov processes, *Theor. Prob. Appl. 3*, 202–246.

Skorohod, A. V. (1965). *Studies in the Theory of Random Processes*, Addison-Wesley, Reading, Mass.

Soong, T. T. (1973). *Random Differential Equations in Science and Engineering*, Academic Press, New York.

Stratonovich, R. L. (1966). A new representation for stochastic integrals and equations, *SIAM J. Cont. 4*, 362–371.

Sussman, H. J. (1978). On the gap between deterministic and stochastic ordinary differential equations, *Ann. Prob. 6*, 19–41.

Takeuchi, K. (1981). Comparison theorems for solutions of stochastic differential equations, *Mem. Fac. Sci. Kyushu Univ. Ser. A 35*, 173–184.

Turelli, M. (1977). Random environments and stochastic calculus, *Theor. Pop. Biol. 12*, 140–178.

Turelli, M. (1978). A reexamination of stability in randomly varying versus deterministic envirnments with comments on the stochastic theory of limiting similarity, *Theory. Pop. Biol. 13*, 244-267.

van Kampen, N. G. (1981). Ito versus Stratonovich, *J. Statist. Phys. 24*, 175–187.

Wagner, W., and E. Platen (1978). Approximation of Ito integral equations, preprint ZIMM, AdW der DDR, Berlin.

Watanabe, S., and T. Yamada (1971). On the uniqueness of solutions of stochastic differential equations II, *J. Math. Kyoto Univ. 11*, 553–563.

Wong, E. (1971). *Stochastic Processes in Information and Dynamical Systems*, McGraw-Hill, New York.

Wong, E., and M. Zakai (1965a). On the relation between ordinary and stochastic differential equations, *Int. J. Engng. Sci. 3*, 213–229.

Wong, E., and M. Zakai (1965b). On the convergence of ordinary integrals to stochastic integrals, *Amer. Math. Statist. 3*, 1560–1564.

Wright, D. J. (1974). The digital simulation of stochastic differential equations, *IEEE Trans. Auto. Cont. 19*, 75–76.

Yamada, T. (1973). On a comparison theorem for solutions of stochastic differential equations, *J. Math. Kyoto Univ. 13*, 497–512.

Yamada, T. (1976). On the approximation of solutions of stochastic differential equations, *Z. Wahrschein. Ver. Geb. 36*, 153–164.

Yamada, T., and S. Watanabe (1971). On the uniqueness of solutions of stochastic differential equations, *J. Math. Kyoto Univ. 11*, 155-167.

Yamato, Y. (1979). Stochastic differential equations and nilpotent Lie algebra, *Z. Wahrschein. Ver. Geb. 47*, 213–229.

Yeh, J. (1973). *Stochastic Processes and the Wiener Integral*, Marcel Dekker, New York.

Yoshizawa, T. (1966). *Stability Theory by Liapunov's Second Method*, Math. Soc. Japan, Tokyo.

Author Index

Subject Index

Accessible boundary (*see* Attainable boundary)

Attainable boundary, 146–148

Attracting boundary, 146–148, 155

 for logistic equation, 170, 182

Backward equation (*see* Kolmogorov equation)

Bellman-Gronwall inequality, 69

Bivariate normal distribution, 14

Borel-Cantelli lemma, 3

Boundary classification, 145–150, 155, 156

 for logistic equation, 167, 171, 182

Brownian motion process (*see* Wiener process)

Carrying capacity, 169, 170, 172

Cauchy-Schwarz inequality, 8

Central limit theorem, 18

Change of variables, 7, 8, 109

Chapman-Kolmogorov equation, 26, 27

Characteristic function, 9

Chebyshev's inequality, 9

Colored Gaussian noise, 121

Competition model, 182

Conditional expectation, 10–14

Conditional probability, 10, 12, 13

Consistency condition, 217

 (*see also* Reducibility condition)

Convergence

 mean square, 15

 of numerical methods, 184, 188, 193, 200, 201

 stochastic, 15